RACISM, NOT RACE

Racism, Not Race

ANSWERS TO FREQUENTLY ASKED QUESTIONS

Joseph L. Graves Jr. and
Alan H. Goodman

Columbia University Press
New York

Columbia University Press
Publishers Since 1893
New York   Chichester, West Sussex
cup.columbia.edu

Library of Congress Cataloging-in-Publication Data
Names: Graves, Joseph L., 1955– author. | Goodman, Alan H., author.
Title: Racism, not race : answers to frequently asked questions /
Joseph L. Graves Jr. and Alan H. Goodman.
Description: New York : Columbia University Press, [2022] |
Includes bibliographical references and index.
Identifiers: LCCN 2021014721 (print) | LCCN 2021014722 (ebook) |
ISBN 9780231200660 (hardback ; alk. paper) | ISBN 9780231553735 (ebook)
Subjects: LCSH: Racism. | Race.
Classification: LCC HT1521 .G678 2022 (print) | LCC HT1521 (ebook) |
DDC 305.8—dc23
LC record available at https://lccn.loc.gov/2021014721
LC ebook record available at https://lccn.loc.gov/2021014722

Columbia University Press books are printed on permanent and durable acid-free paper.

Printed in the United States of America

Cover design: Milenda Nan Ok Lee

For those whose shoulders we stand upon, especially Kazutoshi Mayeda, Beverly Rathcke, and John Vandermeer (JLG); and George J. Armelagos, Franz Boas, and Frederick Douglass (AHG), mentors, professors, and intellectual ancestors who helped us see farther.

CONTENTS

CONTENTS

QUESTIONS

PREFACE

In the spring of 2019, when America was on the brink of a racial crisis, we first discussed collaboratively writing this book. Much has changed and stayed the same in the last two years. COVID-19 became a global pandemic in March 2020; we fought about the best ways to "dampen the curve"; as we write, vaccines are being distributed, and yet, a full year later, we continue to suffer through the biggest health crisis in a century. Black and brown individuals have disproportionately been infected with and died from COVID-19, yet another example of racial inequalities that are everywhere and linked to racism, not biological race.

On May 25, 2020, as we began writing, a white police officer, Derek Chauvin, casually kept his knee on the neck of a Black man, George Floyd, for 9 minutes and 29 seconds. Floyd, who had been pleading for his life, was killed. Among his last words: "please man!" and "Momma!, I'm through." Other officers watched without showing emotion. Protests of police violence against Black men and women spontaneously emerged in thousands of cities throughout the United States and around the world. Signs proclaiming "Black Lives Matter"—as if that were a radical idea—adorned lawns.

We also suffered through the presidency of Donald Trump, who was finally voted out of office on November 3, 2020. He was kicked off Facebook and Twitter because of his stream of hate filled lies. Yet the racism he brought to the surface is still with us. The COVID-19 pandemic finally might be slowing, but it, too, exposed existing racial inequalities. Derek Chauvin has been convicted of murder. Conspiracy theories intersected with beliefs by the so-called neo-Nazi (though we are not sure what is neo about them) and white nationalists, leading to the storming of the U.S. Capitol on January 6, 2021, with Confederate flags and chants to kill members of Congress and the vice president. Much has changed, but racism remains.

American society has never been without the idea of race and the reality of racism, which are linked and part of the founding and foundation of the United States. In 2016, President Barack Obama hosted a national conversation on race. His aim had been to transform our understanding of race from "charged" to "smarter and deeper." And yet, even with all this talk about understanding race and combating racism, we never got close to his hope for smarter and deeper. We never even got to thinking about race as not a thing but, rather, an idea we made real about humans and their differences.

As a presidential candidate, Trump gained the support of racists and white supremacists with his railings against Mexicans and Muslims. Close to a majority of voters approved and elected him later that year. Over the years of his presidency, Trump and his administration set an example of intolerance that has brought into public view racial hatred and quite possibly has deepened racial divisions. The FBI reports that hate crimes, of which the majority are motivated by racism, reached an all-time high in 2018, the last year for which the agency has complete data.[1]

Witness the tragedy in Charlottesville, Virginia, in August 2017. Thousands of self-proclaimed white nationalists descended on the city to "unite the right." Their purpose, they claimed, was to protest the removal of a public statue of Confederate General Robert E. Lee, but what they did was spew racist hate. They chanted, torches in hand, "The Jews will not replace us." Heather Heyer was murdered when white nationalist James Alex Fields Jr. drove his car through a crowd of counter-protesters. Then, on October 27, 2018, another white supremacist planned and carried out the killing of eleven Jews at the Tree of Life synagogue in Pittsburgh. Robert Bowers, the alleged shooter, had targeted Jews because he believed they were aiding immigration. Hate crimes against Jews in the United States reached an all-time high in 2019.[2]

In September 2019, a woman in Atlanta had the "N" word scrawled on her receipt after dining in a Mexican restaurant. Later that same month, two white University of Arizona students brutally assaulted a Black student. White supremacist massacres are on the rise, including the horrific mass shootings like the ones at two mosques in Christchurch, New Zealand, during Friday prayer on March 15, 2019, and another targeting Mexicans at a Walmart in El Paso on August 3, 2019.

If 2019 wasn't bad enough, the horrible killings of nonwhites (specifically Blacks) intensified in 2020. Months before the murder of George Floyd in Minneapolis, Ahmaud Arbery was brutally killed by vigilantes in Georgia while out for a jog, and Breonna Taylor was shot and killed by police in Kentucky as they executed a no-knock warrant on her apartment, apparently in search of a former boyfriend. Jacob Blake was shot in the back seven times by a police officer responding to a domestic violence call in Wisconsin on August 23, 2020.

Meanwhile, less attention-grabbing racial inequalities persist, as represented in infant mortality and deaths from chronic diseases. Black infants die at twice the rate of white infants. Hispanic families have less than a tenth of the wealth of white families. Race is undeniably important from the cradle to the grave and pretty much every age in between. But what, exactly, is this powerful idea called race? Even at the height of this crisis, the vast majority of Americans are left without factual and clear answers to this most fundamental of questions. Many even get anxious about using the right name to designate members of a particular race. Is it Native American or American Indian? Is it African American or Black?

We get stuck on names and drive into a sort of superficial racial conversation gridlock. Fewer follow President Obama's hope and focus on the underlying question of what is race and its connections to racism. As individuals and as a society, we seem to be dazed and confused when it comes to race. The confusion is deep but also something we can solve if we ask the right questions and get clear answers.

Most people who are fighting against racism are doing so with their metaphorical hands tied behind their backs because they are not clear about what race is and what it is not. They might know about the history of slavery and other forms of racism, but they are limited in their knowledge of the history of the *idea of race*. Most important, many do not know that the science of human genetic variation shows with certainty that there are no biological races. Without this knowledge, it is difficult to confront biases that are based on biological and genetic myths about race. Knowledge about what race is and is not is a necessary tool to pull out racism at its roots.

So, how do we win the centuries-long fight against racism and for racial justice? First, we need to ask the right fundamental questions about race. Most people, including socially progressive ones, think race is real, and they are obviously right. Race is real. But *how* is race real? Critically, race is not real in the way most of us have come to think of it: as natural, fixed, and based on biological differences. Beverly Tatum, psychologist and former president of Spelman College, says this view of race is akin to ideological smog that we all breathe every day of our lives.[3] Violent racists breathe this polluted air. We all do. It is time to stop polluting and clean it up.

For centuries, it was assumed that the idea of race was founded on biological differences. But that idea is as wrong as continuing to believe that the earth is flat or the sun revolves around the earth. Disentangling the idea of race from the tangible reality of biological variation allows us to see how this idea of race fuels racism. And it allows us to see that institutions and everyday racism, not minuscule biological differences among races, explains the glaring inequalities in infant mortality, life expectancy, and other critical aspects of life. Understanding what is and is not race can eliminate racial smog at the source.

It may surprise most readers to hear that, like the overwhelming scientific consensus on climate change, scientists and anthropologists have for decades had a clear consensus about what race is and what it's not. Although the idea that biological races exist among humans has been completely dispelled, this advancement in understanding and its implications have somehow failed to influence the way most individuals think of race. Among other causes, this lack of public understanding can be linked to the vested interests of a society that is structured on racism, the inability of some scientists to craft a clear and appealing antiracism narrative, the conservative nature of science, the inability of scientists to let go of the dominant paradigm of human races, and the failure of society to support broad scientific literacy.

Despite the advancement of scientific understanding concerning the relationship of human genetic variation to conceptions of race, the United States (and many other countries) seems to be backsliding toward greater racial misunderstandings and intolerance. Thus, in this work, we hope to impart three important lessons:

1. Racism created the idea of race.
2. The idea of race has real effects. It gives cover to racism.
3. Human genetic variation is real (and quite wonderful) and is absolutely not the same as race.

It's crucial for all of us to know what race is not. Race is not based on biology and genetics. Race is not "in the genes," and it is not the same as human variation. Racial differences in opportunity and outcomes cannot be blamed on genetics.

Unpacking this myth of a link between socially defined races and genetic variation requires an understanding of the relationship among social classification, evolution, and how human genetic variation is globally distributed. Our goal in this book is to clearly show what the powerful idea of race is, how it intimately connects to racism, and how it has been falsely linked to biological variation. Once these lessons are understood, we can dispel many myths about race associated with, for example, health, athletic ability, and intelligence, and present a path for living in a more just and equitable world. Because contemporary myths about race have their roots in the concept's history, we will help readers to add context to what race is and what it is not by stepping into the fascinating history of the idea of race, one akin to an emperor with no clothes (which, incidentally, is the title of Joe's first book on this topic). We then bring our discussion up to the present and look ahead to a more equitable and just future. We recognize that this is no small feat. But in small portions it is easy—and liberating—to understand.

Between the two of us, we've taught and written on race, racism, and human genetic variation for almost a century. Here we provide concise and accessible answers to the questions about race and racism that routinely come up during our conversations with students, educators, policy makers, friends, family, and the public.

This book is for people who want to work toward human equity and those who are nervous about saying the wrong thing. In many cases, the two groups overlap. We hope to encourage readers to have deeper and factually correct conversations concerning race, racism, and human variation. As we discuss in the following pages, this is a conversation that we dare not get wrong.

Throughout the book, readers will encounter different schemes for defining socially designated racial groups. For example, the U.S. Bureau of the Census uses terms such as white, Black or African American, American Indian or Alaska Natives, Asian, and Native Hawaiian or Pacific Islander (see chapter 10). These terms might be familiar to most people but can also be misleading. Some schemes of racial definition use colors (black, brown, red, white, yellow), whereas others utilize variants of nineteenth-century anthropological and fake science categories (Caucasian, Mongoloid, Negroid). Whenever possible we adhere to the principle of parallelism when we describe socially defined racial groups. So, if we say Black, all other groups discussed in that section will be described in the color scheme system. In addition, because we cite examples from primary disciplines (anthropological, biomedical, genomic, historical, and sociological research), material from those citations will use the terms employed by those authors. Throughout this work we provide readers with examples that illustrate the limitations of all of these schemes when they are used to describe human biological variation. Finally, the confusion arising from the various schemes and names and numbers of races is also further evidence that there are no biological races. Furthermore, it is proof of the chameleon-like nature of race, which changes its appearance to fit the needs of those with political power.

Finally, we use technical language because a good many of the terms carry an important precision. Whenever possible, we define the terms so that they are clear to the reader. As well, a number of online glossaries are available, including the National Human Genome Research Institute's Talking Glossary of Genetic Terms (https://www.genome.gov/genetics-glossary), American Anthropological Association's public education project on race (https://understandingrace.org/Glossary), PBS's Evolution Glossary (https://www.pbs.org/wgbh/evolution/library/glossary/index.html), and the Dictionary of Anthropology (http://www.anthrobase.com/Dic/eng/).

ABOUT THE AUTHORS

This is the first time that Joe and Alan have written together. We took the leap because of our mutual goal of increasing public understanding about what race is and is not and how race interrelates with racism and does not explain human variation. Our personal stories, positions, and training are mutually complementary. Alan approached Joe with the idea for a question-and-answer book. Joe had already been considering such a format.

We have lived thousands of miles apart and have led very separate and distinct lives, but we came to share deep concerns and commitments. Each of us grew up in the 1960s and came into our chosen academic fields in the 1970s and '80s. Different paths led us to our present positions as senior academics and scientists. We have always shared a commitment to science and social justice.

Alan's parents lived valiant lives. They were working-class Jews who strived to live decently. Alan's dad was a refrigeration mechanic who loved ideas and hated bigotry, and although he surely faced it, he never talked about anti-Semitism. His mother was silent until near her final days about her own mother's life, including how she escaped to the United States after being brutally tortured, part of a routine Russian pogrom.

Alan, a couple years older than Joe, grew up in Quincy, Massachusetts, a suburb of Boston. In the 1960s, shipbuilding was the main industry there. Quincy was composed mainly of Irish and Italian families. Alan's dad was a World War II veteran and came home to marry, start a family (of five children), and train to become a refrigeration mechanic and a member of the pipefitter's union. Growing up, Alan was well aware of his duality as a working-class kid, along with most of his friends, and a Jewish kid, unlike his Catholic friends. Alan became much more aware of the salience of his white skin privilege when he went to college.

Joe lived racism. His mother and father survived Jim Crow in Virginia. His father landed on Utah Beach in 1944, participated in the battles to liberate France and the low countries, and was in the Arden Forest during the Battle of the Bulge. He helped to build a bridge across the Rhine and was decorated three times. Still, when he came home to America, jobs and other opportunities were denied him due to his race. Joe's mother was "the help," and after she cleaned the homes of wealthy women during the day, she worked in a plastics factory at night. Joe's background made him an unlikely candidate for becoming a scholar in evolutionary biology.

Joe is an evolutionary biologist (the first African American to earn a PhD in this field). His primary research concerns the genetic basis of adaptation. His research training was "pre-adapted" to understanding the nature of biological variation within species. This phenomenon is the basis of all biological conceptions of race. Thus, it was easy for him to see the utter lack of concordance between social characterizations of race and the actual distribution of biological variation within our species. As an African American, Joe has lived experience as a racially subordinated person in the United States. This has informed his understanding of both the conceptual basis and actual practice of racism.

Alan is a biological anthropologist with a deep interest in both the history of racial science and the details of the science. His perspective is that one can critically evaluate racial science on two intersecting levels: in context and on its own terms.

At the contextual level, one can evaluate the reasons a study might have been written and its assumptions. If one is taking time to measure the size of skull differences among races or testing a drug on different races, then one assumes that this question is important and that there is an assumption that race is a meaningful biological category. One can also ask, "What does this study do for us?"

At the scientific level, one takes the study at face value, assumes that the assumptions are justified, and then evaluates the study purely as a piece of science. Questions one might ask include whether the methods are sound and if the conclusions follow from the results.

Alan learned about the myths of race as an undergraduate and early in graduate school. Back in the early 1970s, he assumed that the destructive idea of race—an idea used against Jews and Black and brown peoples—would go away. Don't flawed explanations, like phlogiston and the homunculus, go on the scrap heap of dead scientific ideas when proven wrong? But he did not realize the power that race held. And so he began to study race science in the early 1990s.

ACKNOWLEDGMENTS

Any book is a project of both solitary efforts and a network of supporters. We would like first to acknowledge those who have helped envision this book and bring it to press. Beth Vessel and Callie Deitrick found this book a supportive home during the pandemic with Columbia University Press. That was no small feat! Fiona Marks, Hampshire College undergraduate, helped with references and proofreading. Eric Schwartz, editorial director, and his staff at Columbia University Press have been thoroughly responsive and have greatly improved the accessibility of the book.

Joe would like to acknowledge his students from his evolution, genetics, and anthropology courses. Their questions helped clarify his thinking. Joe dedicates this book to those whose shoulders he has stood upon, especially Beverly Rathcke (1945–2011), who taught him how to be critical of received scientific knowledge; John Vandermeer, who taught him that biology could be a social weapon; and Kazutoshi Mayeda (1928–2008), who pushed him to better understand human genetics.

Alan also wishes to acknowledge the colleagues who have been in front of and beside him in his efforts to teach a nonracial approach to human variation and the compatibility of good science and social justice. These include, but are certainly not limited to, Michael Blakey, Joseph Jones, and Yolanda Moses and the many colleagues who worked with him on the AAA public education project on race (understandingrace.org).

Alan also wishes to express his thanks to those who helped him think in creative ways about how to express scientific information and ideas while working together on the PBS documentary *Race: The Power of an Illusion* and the AAA traveling exhibit on race and, most recently, the *RaceGen* list server. These include Robert Garfield, Joanne Jones-Rizzi, Larry Adelman, Llew Smith, and Christine Sommers. Many other colleagues have supported him throughout his efforts to educate people about race, racism, and human variation, including Clarence Gravlee, Ricardo Santos, Faye Harrison, Leith Mullings, Audrey Smedley, Agustin Fuentes, Ken Kidd, Brooke Thomas, Ann Mourning, Lynn Morgan, Thomas Leatherman, Troy Duster, Charmaine Royal, Alondra Nelson, Nancy Krieger, Deborah Bolnick, Jonathan Kahn, Evelynn Hammonds, and Jon Marks. George Armelagos taught him that science could be fun, fulfilling, and relevant.

Alan also wishes to acknowledge his parents, whose humility and bravery taught him more than they will ever know. It has been a pleasure for Alan to write this book with Joe Graves. He is humbled and honored. This book would absolutely not be possible without the love and support of Alan's wife and daughter, Chaia Heller and Ruby Heller-Goodman.

RACISM, NOT RACE

WHAT ARE RACE, RACISM, AND
HUMAN VARIATION?

A few years ago, a student of African descent came into Joe's office worried that she had scleroderma, a chronic connective tissue disease that is generally thought to be part of the family of autoimmune rheumatic diseases. She sounded confused and told Joe that her doctor would not diagnose her disease as such. She recalls her doctor saying something about "Black people not getting scleroderma."

Marta, a fifty-something-year-old colleague of Alan's, has Native American, European, and African ancestry. Like two-thirds of the people in the world and all other mammals, she is lactose intolerant because she does not produce sufficient lactase, the enzyme that digests lactose. Thus, she avoids consuming dairy products. Because of her restricted diet, Marta feared that her calcium intake was low and that it might put her at risk for osteoporosis. She raised her concerns with her primary care physician and asked if he would order her a bone density test. Her doctor responded that she didn't need one because she is Black, and Blacks do not get osteoporosis.

Despite their intensive medical training, these two doctors made a common error. They made judgments about their patients' race based on their "looking Black." Black, that vaguely defined group, is culturally real and visible in our society and to the medical profession. "Black" or "African American" is what we call the members of a group who were not long ago referred to by other terms, such as "Negro." And who is Black, or of any other race, has always been poorly defined. But that Black is socially salient, an indexed group, is certain. Medical textbooks are chock full of statements about different rates of disease by race: Black is a "risk group." Even the thoroughly outdated "negroid race" is still a medical search term.

Regarding scleroderma, a pamphlet from the Scleroderma Foundation proclaims that "race and ethnicity may influence the risk of getting scleroderma," and the website Sclero.org states, "Factors that may incline a person to develop scleroderma include race, ethnicity, and geography."[1] Regarding osteoporosis, echoing lots of medical textbooks, the popular website WebMD states, "Research shows that Caucasian and Asian women are more likely to develop osteoporosis than women of other ethnic backgrounds. Hip fractures are also twice as likely to happen in Caucasian women as in African-American women."[2]

So, as Alan's friend and Joe's student looked Black to their doctors, the physicians decided to not offer a diagnosis and not treat the women because they thought that the risk of disease was diminished because the patients were Black. These two medical care errors take us from social classifications of race—specifically, "Black" in these cases—to genetic differences among individuals—what we will refer to later as the amazingly interesting global structure of human genetic variation—and its significance for disease risk.

These physicians' faulty assumption that human genetic variation is patterned along socially defined race or color lines is deeply embedded in how nearly everyone, including the vast majority of doctors, think about race. We see similar slippages in the day-to-day events and decisions made in hospitals, schools, banks, courts, and other institutions. A system of ideas that links race and human genetic variation underlies the fraught interactions we witness every day. This is a huge scientific error, because social classifications of race change over time and place. They are unstable. And all known social classifications fail to map onto genetic variation. This easy conflation of social race with genetic variation is the most consequential error of how we have thought and continue to think about race.

In this introductory chapter, we do an initial unraveling of the problematic connections between the social categorization of race and the realities of human genetic variation. We explore how the mythic idea of biological race within our species continues to give ideological cover to racism. So, what better way to start than with the big three questions we encounter in the classes we teach, in the presentations we give, and in our own lives:

What is race?
What is human genetic variation?
What is racism?

We will also discuss how these concepts have been falsely connected and how they should be connected.

WHAT IS RACE?

Here is our short definition: race is a worldview and social classification that divides humans into groups based on their appearance and assumed ancestry, and that has been used to establish social hierarchies. Our definition is that of *social race*. Race is a social classification based on assumptions about ancestry and appearance. Socially defined races are unstable categories in that definitions, names, and color lines change. Nonetheless, social race is very real, as exemplified by differences in wealth and health.

Many still think that the way we classify race in humans is fundamentally biological. However, that race, what we call *biological race*, does not exist in our species. It is a long-standing myth that provides cover for racism. But race is real as a social category.

RACE IS REAL AND NOT REAL? HUH?

Nearly everyone is confused about what race is and what it isn't. Parents, pollsters, principals, and politicians are confused. People of all skin colors, ethnicities, classes, and religions are confused. If you are, too, you're not alone.

Part of the problem comes from the fact that what everyone, including scientists, thinks race is has changed over time. Moreover, we have never had a clear and broadly accepted definition or even known what should be part of the definition.

A first dip into history. Carl Linnaeus (1707–1778) was one of the first in a long line of scientists to try to define and classify races. He was a Swedish naturalist who lived in Uppsala, just north of Stockholm, and is considered the father of modern taxonomy, the science of classification. Linnaeus invented the modern binomial nomenclature that separated all living things into species and genus, and then lumped species and genus into broader categories. For example, we now agree that we are all part of the genus and species *Homo sapiens* and are part of the order Primates, along with chimpanzees, gorillas, orangutans, monkeys, and prosimians.

Linnaeus traveled at the beginning of his career but mostly relied on others to send him descriptions of plants and animals (and better yet, samples when possible). Linnaeus believed that he was put on earth to classify God's creations and took as his modest motto, *Deus creavit, Linnaeus disposuit* ("God created and Linnaeus organized"). His creation of the binomial system of classification was quickly and widely regarded as a great scientific advance. Philosopher-naturalist Jean-Jacques Rousseau said, "I know of no man greater on Earth."[3]

Linnaeus wrote a number of editions of *Systema Naturae* (*Systems of Nature*).[4] His first, published in 1735, and his tenth, released in 1758, expanded his thoughts on the taxonomy of humans. He called humans *Homo sapiens*, as we still do, and divided us into varieties based on the information he received from travelers. He included categories of monstrous and wild men that later could not be confirmed as human. He also included four continental and scientific-sounding varieties of human: *Europaeus, Asiaticus, Africanus*, and *Americanus*. Linnaeus described them all by color, temperament, and governance. For example, *Europeaus* was white, clever, and ruled by law, and *Americanus* was red, free, and ruled by custom.

These ideas about each variety morphed into what eighteenth-century naturalists came to think of as subspecies and races. That idea is based on two underlying notions: stability of type and hierarchy. These notions of stability and hierarchy continue to undergird racist ideologies today and are elements of the pervasive racial smog that we all breathe.

Working within the Platonic system of imagining that all things, living or inanimate, have distinct attributes, Linnaeus and other naturalists considered each race as similar to other things, each with its own essences and specific traits. The idea of race was founded on the notion of an unchanging earth created by God, so that what is now has always been.

The aspect of hierarchy is traced to the doctrine of a great chain of being, which began with the Greeks and was later adopted by Christians. All creatures were arranged in order of their relative closeness to God. Linnaeus's job was simply to describe and put these things in order. God created, and Linnaeus classified.

Importantly, the idea of race is interwoven with colonization and slavery. Vilifying reports about newly encountered peoples from colonizers and slavers often made their way into Linnaeus's classification. And his classification looped back and lent scientific support for slavery and colonization. Having bent science to their side, the aristocrats of Europe adopted a view of slavery and colonization as just and inevitable. Africans were seen as inherently incapable of governing themselves, and Native Americans were not God's children. That formulation of race was invented and quickly accepted because it provided a scientific cover for colonization, plundering, and slavery. The institutional racism of slavery and colonization gave vitality to the essentialist biological race concept. More briefly: racism made race.

Of course, in the end, the science of putting humans into discrete races failed because humans cannot be divided into convenient color categories or biological races (see chapter 3 for details). But that critical fact has done little to derail the essentialist view of human variation that still fuels virulent racists and unfortunately is still the primary way in which most people understand race and human biological variation. It was behind the enslavement of fifteen million Africans. It was behind the murder of six million Jews in Europe. It was behind the murder of eleven individuals, including a Holocaust survivor, at the Tree of Life Synagogue in Pittsburgh in 2018; the Christchurch mosque murders of fifty-one people in March 2019; the murder of twenty-two people, mostly Mexican Americans, at a Walmart in El Paso in August 2019; and the violent storming of the U.S. Capitol on January 6, 2021.

These views of race as biological, innate, and hierarchical are not limited to eighteenth-century plantation owners, Nazis, and twenty-first-century white supremacists. Most people believe that something innately genetic makes each race unique. Indeed, a recent survey of more than sixteen hundred Americans demonstrated that the vast majority of individuals felt that race was both a social category and based on biological attributes.[5]

Thus, nearly six decades after the passage of the Civil Rights Act of 1964, views continue that align with eighteenth-century pre-Darwinian and mythical thoughts that racial differences are natural and that some races are superior to others. Laws can change suddenly, as with the Civil Rights Act, but it takes much longer to change worldviews, especially those that support the status quo.

In a democratic society, changing laws is a long and complex process. Serious struggle ended slavery and led to the passage of the Civil Rights Act a century later. Yet, as hard as it is to change laws, it is even harder to change how citizens think and what we all believe in. And beliefs are difficult to change when they come with deep investments. Indeed, the growing polarization we are observing in American society is associated with the fact that people hold to different systems of what they believe to be the truth.

So, race as a worldview and social classification divides humans into groups based on assumptions about their appearance and ancestry. It came about as a justification for colonization, appropriation of land and goods, and enslavement. Before scientists got involved, race began as a folk idea about differences between and fears of the "other." Racial thinking was likely to be a form of ethnocentrism, the belief that one's society and culture is superior to those of others.

What's unique about race is that scientists made race real through their pseudoscientific studies purporting to show that races were biological units and that differences between races were immutably based on biological distinctions. This process of trying to make something seem real by constant use—especially when supported by those with authority, such as scientists—is called *reification*. Biological race, even though it is fiction, became reified. It came to be seen as unquestionable and real. Biological races are a myth. Throw them on the scrap heap of dead scientific ideas. But social race lives on.

HOW MANY RACES ARE THERE?

If by "race" we mean clear and logical divisions in the human species based on evolved genetic differences, then there are no races. Human beings are simply not divisible into unambiguous biological races. Why this is true is explained in chapters 3 and 9.

If, on the other hand, we see race as a social classification, then the answer to "How many races are there?" is, "It depends." It depends on one's goals. For example, the number of races in the Brazilian census is different from the name and number of races in the U.S. census. This is due to different histories and different configurations of social and political importance of the race concept in each country. Moreover, the number of races and how we refer to them also changes over time. For example, in the U.S. Census of 1790, only five categories were enumerated: Indians, free white males, free white females, all other free persons, and slaves. In the 1850 census, the race categories were Indian, white, black, and mulatto. What's most striking at the time when slavery was being challenged is the addition of "mulatto"—a person of both white and black parentage. In 2000, the following categories were recognized: white, Negro or black, Japanese, Chinese, Filipino, Korean, Vietnamese, American Indian, Asian Indian, Hawaiian, Guamanian, Samoan, Eskimo, Aleut, and Other (specify).

In Brazil, efforts have been made to track individuals by skin color, and this reflects the fact that admixture (reproduction between people of different geographic ancestries) has been much greater in that country than in the United States. Brazilian categories today include *brancos* (white), *pretos* (black), and *pardos* (mixed or multicultural) and are designed to capture degrees of mixing among Europeans, Africans, and indigenous South Americans.

The U.S. census and the Brazilian census are but two examples highlighting the fact that official racial classifications are unstable. They bend to meet the needs of those in power. In addition, racial terms in common use are even more fluid. The assumed definitions of the categories also change. The criteria for inclusion of an individual in a race is both fluid and ill-defined. This is particularly problematic in scientific studies, in which reliability, the ability of different researchers and the same researcher to classify individuals in the same way over time, is paramount. Without reliability, race science is a version of junk in, junk out.

Witness the racial category of European or white. In 1939, Carleton Coon, distinguished professor of anthropology at Harvard, wrote in *The Races of Europe* that there were eighteen or so European racial types. He mapped out and described the biological characteristics of groups such as Lappish, Nordic, Dinaric, Alpine, Baltic, Upper Paleolithic Survivors in Ireland, Long-Faced Mediterraneans, Armenoids, and Jews. After the horrors of the Holocaust, many of the groups that were not fully considered to be white by American Anglo-Saxons began to be more fully accepted as white. Yale historian Mathew Frye Jacobson refers to these groups as whites of a different color.[6]

Race is a flexible classification because the salience of social categories changes over time and place and with different needs. In other words, race is a chameleon; it changes to fit different social and political environments. That makes sense for a social concept. However, this instability of racial classification makes it impossible to generalize. That flexibility is also not acceptable for a scientific classification, as all classifications depend on consistency of reliability classification. Finally, social races do not map onto any sort of genetic divisions.

WHAT DO GENETICISTS MEAN BY THE STRUCTURE OF HUMAN VARIATION?

A key question, maybe *the* key question, is whether the concept of race describes or explains biological differences within our species. To get to that question, here we will take a first step into the world of science and come to understand what scientists mean by the term "structure of human variation" and then how variation comes about.[7]

People are unique. To understand the reality of biological variation, all you have to do is look at the people around you in a room or on a street. No two individuals look the same. Even genetically identical twins have subtle differences that parents discover and use to tell them apart. Indeed, we have learned that identical twins differ in non-nucleotide-based genetic variation (*epigenetics*). These epigenetic differences can lead to sometimes large differences in traits, such as twins' predisposition to and severity of disease.[8] A recent example is shown by identical twins who developed dramatically different levels of illness when infected by COVID-19.[9]

Humans are a highly visually aware species, and we home in on the incredible variety of observable appearances in skin tone; hair length and style; nose, eye, mouth, and ear shapes; height and body shapes; and even how a person stands and moves. There are also many differences that require special instruments to uncover, such as blood pressure and blood type, along with a host of other physiological, metabolic, and genetic traits. Evolutionary geneticists study how variations change over time, and population geneticists study how variations are patterned or structured within individuals and among individuals and groups.

By "structure of human variation" we mean two interrelated things. The first is the pattern of variation among traits within individuals. For example, height and weight are strongly related, or correlated, whereas height and blood types are not. The second meaning refers to the pattern of differences and similarities among humans at three levels: among individuals within a group, among groups, and among purported races. One issue we encounter is that what is meant by group—or, biologically speaking, a population—and especially by race, is not consistently defined. As noted earlier, groups and especially races are poorly reproduced. A group, for example, could be an ethnicity or a country. We know from the previous discussion that these groups and populations are socially defined and that the criteria for who is in and out change through time. Definitions and group memberships are unstable.

That caveat aside, the structure of variation is, in a sense, about levels of variation: within an individual, among individuals within a group, among groups, and among purported races and across the globe. A key test of the salience of race is the amount of variation that exists among socially defined races. As we will illustrate, it is shockingly small.

There is another distinction we want to introduce before moving on: genotype versus phenotype. A phenotype is a fancy word for any observable trait or characteristic. This includes visible and easily measurable traits, such as height, and also more complex characteristics, like personality and musical ability. Are you a dog person? Do you like games? Are your eyes glazing over? These are all phenotypes.

Some phenotypes are strongly determined by development and aspects of the environment, and others are fundamentally due to the genes you were born with. And the vast majority are a complex mix of genotypes, environments, and interactions of genotypes and environments.

Genotype refers to the composition of your deoxyribonucleic acid (DNA). You've probably heard that our genomes are composed of DNA, a long molecule that has the shape of a double helix. DNA itself is a string of nifty chemicals called *nucleotides* that point inward and are held together by strong hydrophilic (water-loving) forces. The four nucleotides of DNA are adenine, thymine, cytosine, and guanine, abbreviated to A, T, C, and G. The A's and T's are always paired, and the C's and G's are always together.

The full human genome contains about 3.3 billion nucleotide letters. Some of the genome codes for amino acids: a sequence of three nucleotides in a row (triplets) corresponds to one of the twenty-three naturally occurring amino acids. The amino acids are assembled into proteins. Some proteins make crucial structures such as collagen, and others are enzymes and catalysts that promote vital biochemical reactions that keep us going. Proteins are absolutely critical to life and thus so are the amino acids that make up the sequential structure of proteins and so are the nucleotides that code for the amino acids. But of course, life is also more than nucleotides, amino acids and proteins.

Let's briefly consider height as an illustration of the structure of a phenotypic variation. Height varies among individuals within a group, and the average height differs slightly among groups and has varied over time. These are all measurable facts. We do not know much about height variation among socially defined races, because the groups and individuals in each socially defined race change. For example, in America, "Black" is culturally defined as anyone with detectable African ancestry; in Puerto Rico, the term denotes anyone who has no European ancestry; and in Britain, it can include anyone who isn't a European.

Individual height is highly heritable (an estimated 80 percent determined by genes). More than 180 genetic loci have been found to have significant effects, and thousands more are known to have some effect on height. At a more complex level, height is influenced by environmental factors such as early life nutrition and pollution. While we do not know the exact cause of the changes, in recent years, the Dutch have far surpassed Americans in their average height, and the children of first-generation Koreans now have average heights equivalent to those of Americans, even though the parents of that generation had average height less than that of Americans.

What about variation in genotypes? First, all human beings are identical at about 99.9 percent of the DNA nucleotides. There are differences at an average of only about one of every thousand letters.

Consider these random changes in a passage from Harriet Beecher Stowe's *Uncle Tom's Cabin*. The first three passages show single-letter changes (mutations) in different words. The fourth shows a deletion of an entire word, "old." Note that the fourth sentence still makes sense despite the loss of the word; if the word "sobbing" had been lost, the sentence would make less sense.

Passage 1: "O, now, don't—don't, hy boy!" said the old man, almost sobbing as he spoke.

Passage 2: "A, now, don't—don't, my boy!" said the old man, almost sobbing as he spoke.

Passage 3: O, now, don't—don't, my boy!" said the old man, almost sobbing as he spjke.

Passage 4: O, now, don't—don't, my boy!" said the man, almost sobbing as he spoke.

Now let's move this analogy from a book to the human genome. As noted earlier, our genome contains approximately 3.3 billion nucleotides. Of these, about 10 million differ between any two individuals, thus making each individual unique. Below is an example of a random change in a DNA sequence. The sequences of the three individuals show single changes relative to one another. Given that each set of three nucleotides (or triplets) codes for amino acids in the protein resulting from the sequence, we have shown the amino acids that would result from each sequence.

Individual 1 DNA: ATG GG**A** TTC AGA **GGG** CCG TCA TAG
Amino acids: methionine, glycine, phenylalanine, arginine, glycine, pro-
line, serine, stop code.
Individual 2 DNA: ATG GG**C** TTC AGA GGG CCG TCA TAG
Amino acids: methionine, glycine, phenylalanine, arginine, glycine, pro-
line, serine, stop code.
Individual 3 DNA: ATG GGA TTC AGA GG**T** CCG TCA TAG
Amino acids: methionine, glycine, phenylalanine, arginine, glycine,
proline, serine, stop code.
Individual 4 DNA: ATG GGA TTC AGA ___ CCG TCA TAG
Amino acids: methionine, glycine, phenylalanine, arginine, proline,
serine, stop code.

The DNA sequence change between individuals 1 and 2 does not change
the amino acid sequence. This is because the triplet code is redundant with
64 possible codes for only 23 naturally occurring amino acids, along with
one start and three stop codes. Notice that the changes between the first
three individuals' DNA sequences did not change the resulting amino acid
sequence. However, the fourth individual is missing an entire triplet, resulting in the loss of the amino acid glycine in the final protein sequence. The
loss or change of an amino acid in a protein often has severe consequences,
as in the case of the mutation that causes sickle cell anemia, where glutamic
acid is replaced by valine.

Recognizing that individuals differ in their DNA also raises the question
of whether those differences are associated with membership in a "race." In
other words, could the ~3.3 million nucleotide differences be used to neatly
define racial groups within our species? It turns out that the answer is "no."
A quick way to understand this is to break down the approximately 10 million nucleotide differences by geography: worldwide, by continent (Africa,
Americas, Asia, Europe, and Oceania), and by regions within the continents
(see figure 0.1). Genetic variations are most often shared throughout the
world. Others are specific to local environmental conditions. For example,
genes associated with resistance to malaria (e.g., blood group O and glu-
cose six phosphate deficiency) occur in higher frequency in zones where
malaria transmission is still common (the tropics, including sub-Saharan

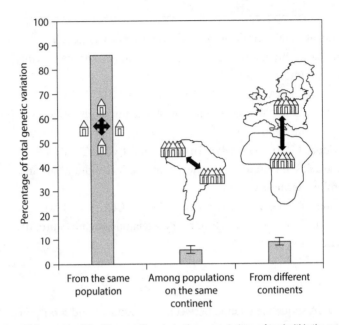

FIGURE 0.1. Of the roughly ~3.3 million genetic variants, the vast majority are found within the same local group or population, and a small percentage is found among populations within a continent and among continents. Moreover, the variation among populations and continents is attributable to geographic distance, not race. *Source*: Courtesy of George Armelagos.

Africa, the Middle East, and India). Genes associated with lactose persistence occur at higher frequencies in populations that domesticated cattle (northern Europeans and East African Masai). There are also genes that help populations survive at high altitude (such as observed in Ethiopians, Tibetans, and Andeans).

Finally, for an example of pattern of variation among traits within individuals, let's consider height, weight, and hair color. Traits having to do with color tend to be correlated. Similarly, traits having to do with size, such as height and weight, also tend to correlate. However, size traits are not correlated with color traits, and color and size traits are generally not correlated with any other traits. The pattern of independence is the factual rule, and it is important. It should cause us to exercise great caution about extrapolating from a single trait, such as skin color or, really, any single trait.

In summary, variation includes genotypes and phenotypes. Although we are all unique, and we certainly look unique, genetically, humans are rather homogeneous. Most of that tiny genetic variation is found within any group.

The largest amount of our genetic variation is cosmopolitan: it has spread throughout the world.

HOW DOES HUMAN VARIATION COME ABOUT?

Variation is amazing and amazingly important. So, how did we get to the pattern of variation seen today? The short answer is that variation evolved. Human variation came about through evolution.

Charles Darwin was the first to clearly demonstrate, in *On the Origin of Species*,[10] that evolution by natural selection results from the existence of variation within species. Some variants lead to greater survival and reproductive success and are therefore selected over other variants. For example, modern humans originally evolved in the tropics. After losing the protection of a hairy body that our ancestors surely shared with their primate cousins, these first humans almost certainly evolved dark skin to better protect them from damage to folic acid, one of the B vitamins. Folic acid, also known as *folate*, is a compound that is essential to the successful completion of pregnancy (live birth, reducing birth defects). Exposure to ultraviolet radiation from sunlight can damage folate, and the melanin in darker skin provides protection.[11]

Darwin had a rudimentary idea about the origin of variation in living things, particularly the importance of variation for adaptation. Today, scientists have a much deeper and broader understanding of how variation comes into existence and evolves. The variation in all biological traits in any group of organisms (including humans) is determined by the complex interaction of three major sources: genes, environment, and chance. For purposes of this book, we're not interested in all biological variation, such as that associated with sex or age. Rather, we're interested in variations among groups and regions that are often associated with socially defined race. How did humans who lived and evolved in different locations come to be different from one another?

Because our species began in sub-Saharan Africa, Africans account for the greatest amount of genetic variation within our species. The rate of mutation in the human genome is constant, so the more time has passed, the greater the number of mutations and accumulation of new genetic variants. Conversely, when smaller groups of humans left Africa and traveled around the world, these founder groups left behind some of the genetic variation that was in the larger population. As a result, the group that migrated lost some genetic variants.

Consider a deck of playing cards as a model of the genetic variants in a large population. If you deal out all the cards, you will find an equal distribution of spades, clubs, diamonds, and hearts. If you deal only four cards, you are highly likely to deal something like two hearts, one club, and one spade. The diamonds have been lost by chance simply because of the small number of cards dealt. If you try this a second time, you might deal two diamonds, one heart, and one club; this time, the spades have been lost. And so on.

This evolutionary process is called *genetic drift* and refers to a random change in allele frequencies due to small population size. An allele is an alternative form of a gene, such as in the case of brown, blue, and green eye color. A consequence of genetic drift is that the farther people migrated from Africa, the more genetic variation (alleles) they lost.[12] But despite the loss of some genetic variants, and the introduction of new variants through mutation during the movement of people to different parts of the globe, at the end of the day, genetically, we all remain Africans.

It is also important to understand that our geographical entities are human creations as well. Australia, as a separate land mass, is somewhat zoologically distinct, whereas Asia and Europe are connected and not so distinct from each other. For example, the best way to understand the distribution of the hemoglobin S (HbS) genetic variant is not by which continents it is found on but by its association with ecological conditions such as stagnant water, which provided a breeding ground for the mosquito that carries the malaria parasite. Individuals who are heterozygous or carriers (one copy of the HbS and one copy of the HbA allele) are resistant to malaria and do not display sickle cell anemia. Individuals without a HbS allele have elevated death rates due to malaria. Individuals with two HbS alleles have elevated death rates due to sickle cell anemia. Thus, in malaria zones, it is best to be a carrier.

Early agriculturalists came in contact with and actually expanded the breeding grounds of malaria on several continents and regions, including west central Africa, the Middle East, the Mediterranean, and the Indian subcontinent. As a result, the frequency of the HbS allele in Nigeria (western Africa) is 0.14, in Saudi Arabia (Arabian Peninsula) it is 0.12, and in Greece (Mediterranean) it is 0.12. Conversely, the frequency of this allele in South Africa (southern Africa) is only 0.02. The difference between the former and latter countries is that the former all have high levels of malaria transmission, whereas there is virtually none in South Africa. The bottom line: continent or race does not explain sickle cell trait, but evolution does beautifully.

Finally, human variation in physical appearance and other phenotypes is also due to how genetics is expressed under different conditions. The environment in which an organism lives always influences the expression of its genes. Picture a tree that lives on a windy slope. The tree grows in a tangled and twisted form because of the constant pressure of the wind. If one takes the seeds of such trees and plants them in a calm environment, the trees will grow to have a more typical, straight trunk.

Throughout human history, people have lived in unique, varied, and often unequal environments. Even in modern times, the probability that a person will die (at any age) in the United States is strongly stratified by socioeconomic environment. A person who lives in a household that brings in less than $15,000 a year is over three times more likely to die than a person who lives in a family whose income is greater than $70,000 annually.[13]

In the United States, socioeconomic status is strongly stratified by socially defined race, so that compared with people of European descent, racial minorities experience more environmental insults, such as heavy-metal pollution and pesticide exposure, during their life spans.[14] The take-home point is that we cannot assume that differences in health or other physical traits (the phenotype) are due to differences in genes and genomes (the genotype). This confusion is one of the most nefarious parts of race science—the assumption that the differences we see in wealth, health, and other measures are attributed to genetic differences when, in fact, they are due to environmental differences.

In summary, genetic variation arises from mutations. Some of these are beneficial, such as in the example of the sickle cell trait in the presence of endemic malaria. Other mutations (or variants) are neutral and simply carried as part of the genome without functional consequences. These variants can change in frequency through time, distance, and random events. Finally, much of what we see, such as differences in life spans, are nongenetic variations that are due to environmental conditions.

IF RACE ISN'T BIOLOGICAL, HOW DO YOU EXPLAIN THE DIFFERENCES WE SEE?

Examples such as the sickle cell allele and skin color (see chapter 3) demonstrate that geographically based genetic variation exists within the human species. However, the existence of these variations should not be confused with socially defined races. Complex forces—ancestry, natural and sexual selection, genetic drift, and environmental effects—alter the physical appearance and other attributes of various populations of human beings. However, it's worth repeating that none of these differences, alone or combined, scientifically justifies the partitioning of people into racial groups. Nor does the underlying genetic variation encoding the physical differences we observe in people justify the clustering of humans into biological races. Both phenotypic and genetic variation is found within and among human groups—but that variation is not racial. It is geographic and is explained by evolution and history. Variation exists. It is wonderful and ripe for exploration. But it is not race.

WHAT IS A SOCIAL CONSTRUCTION?

You might have heard people say that race is "just" a social construct and therefore is not "real." But equating a social construction with not being real is a mistake. A social construction becomes real in that it has real effects.

Our world is socially constructed. How we see our world, our reality, is the result of how our society trains us to see that world. For example, we are a very visual species and see skin color as telling a good deal about an individual. That's not because skin color is inherently more important. It is because skin color is readily seen and has become imbued with cultural meaning.

Anthropologists often study how cultures and subcultures view the world in different ways. A ready example includes attitudes toward race. Within the broad culture of the United States, white people are more likely to see race as biologically based but believe that great progress has been made in reducing racism. In contrast, African Americans and other racial and ethnic minorities are less likely to see race as biologically based, and they are far more pessimistic about progress in race relations.[15]

In Brazil, it is said that money "whitens." Race and class are more tightly wound in the Brazilian compared with the American consciousness. In Europe, anti-Semitism is more closely equated with racism. In medieval Europe, Jews were treated as a separate race of humans. They were thought to have tails, and males were thought to menstruate. Pogroms (organized massacres) against Jews occurred consistently throughout European history. In the Jewish quarter of Paris, the Marais, one frequently encounters plaques dedicated to places where Jewish families once lived who were sent to concentration camps and their deaths. The legacy of European anti-Semitism lives in these places. The difference in how anti-Semitism is understood in

America and Europe results from the fact that people see the world differently based on personal experiences. Our social brains train our eyes to see significance.

Social constructions make our reality. Race might not be based on biology, but, paradoxically, dividing individuals into races has biological effects. Individuals whom our society classifies as Black or brown are treated differently. They have different barriers to achieving a quality education, being hired in meaningful employment, and living in a safe environment. Therefore, as a result, they are exposed to more stress and greater health problems than people who are classified as white. In the United States, the chance that a baby will die before its first birthday is over two times greater if its mother is classified as Black versus white. And the average life expectancy of whites in 2016 approached seventy-nine years, which gives them an extra four years of life compared with Native Americans and Blacks.[16] Social constructions have real consequences for our bodies and our lives.

Another way of understanding social construction is to begin by recognizing that all knowledge is socially constructed. Science operates from the proposition that objective reality exists, and scientists attempt to understand that reality through the use of scientific methods. The scientific method is the most reliable intellectual program to understand how nature works. Among its successes are outlining the rules of motion (by Newton and Einstein), the understanding of the conservation of matter (Lavoisier), cell theory (Schwann), the significance of the cell surface (Just), the origin of species by the means of natural selection (Darwin), and the nature of genetic material itself (Watson, Crick, and Franklin).

The degree to which the people who live in modern industrialized societies enjoy relative ease and comfort is partly the result of the fact that science works. Today, fewer people die of the fatal synergy of malnutrition and infection compared with just decades ago. The worldwide rate of infant mortality has dropped from about one hundred fifty per thousand in the 1950s to under thirty today. At the same time, average worldwide life expectancy has increased from under fifty to over seventy years.

The greatest strength of science is that it requires skeptical analysis and the testing of hypotheses, and therefore it must be self-correcting. However, science is conducted by human beings, and humans always exist in societies. Therefore, as all human beings operate using various levels of flawed thinking, societal biases often enter into the enterprise of science. What's wrong with socially defined race schemes is that they were produced without proper adherence to the scientific method. This deviation from the scientific method was a human flaw, driven by the realization that the idea of race provided justification for the existing sociopolitical hierarchy.[17]

We are left to practice science as best we can, and that includes trying to be clear about our biases. That isn't easy. Scientists are not generally trained to recognize their own biases, which are like the air we breathe.

In hindsight, we can recognize the biases of American and European scientists who tried to justify slavery by making racial hierarchies. And even though seeing one's own biases is difficult, we, too, have our biases. Martin Wobst, one of Alan's graduate school professors, was fond of saying that seeing one's own bias is like jumping over one's shadow. Culture makes and shapes reality.

WHAT IS RACISM?

Racism is systematic discrimination by a powerful individual or institution against individuals based on their perceived membership in a socially defined racial group. Only governments, corporations, social groups, and individuals who have collective political and social power can implement racism. Institutional racism, a predominant form of racism, is alive and well in American society and can be observed in a variety of aspects of our lives, such as discrimination in housing, employment, and income; exposure to toxic wastes; likelihood of incarceration; and death by police intervention. Furthermore, globalization has allowed American-style racism to spread and combine with indigenous racisms to degrade the lives of even more people.

It's important to distinguish bigotry from racism. Anyone can be a bigot— that is, anyone can have an irrational hatred for individuals who belong to a different socially defined racial group. Racism, on the other hand, requires having, and using, social power.[18]

Racism as a worldwide phenomenon dates to the period of the European voyages of colonization. There is no evidence that racism (or the concept of race) existed in the ancient world (~1200 BCE–400 CE). Societies displayed various sorts of ethnocentrism; for example, the Greeks considered all non-Greeks as barbarians. Generally, all cultures thought that their people were somehow better than others with whom they came into contact. Slavery in the ancient world was not racial; rather, it was associated with conquest during war and with poverty. For example, in the Roman Empire, poor people often left at the city dump babies that they couldn't afford to take care of. Slave traders were aware of this practice and congregated at the dump to collect these unfortunate infants.[19]

During the Crusades, the slave trade was balanced between Islamic people entering the Christian trade and Christians entering the Islamic trade. The transatlantic slave trade was unique in the history of humanity in that it targets a specific group based on its geographic location and reliance on social definitions of race. Africans of widely different ethnicities, language groups, and cultures began to be lumped together as a race. European philosophers, such as Giordano Bruno (1548–1600), thought that this group of people had been created separately from Adam and Eve. Others, such as Comte de Buffon (1707–1788), felt that all humans were created from Adam and Eve. Buffon believed that the environment created the different aspects of human physiology, such as skin color and stature. He thought that a savage tribe captured and brought to Europe, fed on European food, would gradually become civilized and their skin would whiten.

The ideology of race always supports those in power. So, it is not surprising that variants of this ideology spread and now it is not solely the province of individuals of European descent. The Japanese imperial class saw the Japanese as the "master race" of Asia and used this ideology to justify its conquest and genocide against other East Asians during the Asia-Pacific War (1933–45).[20] Another horrific example of this thinking was the Rwandan genocide of the 1990s.[21]

Unfortunately, there are several examples in modern history of what happens when racist ideology becomes the policy of the state. When the subject is broached with Americans, most think of Nazi Germany. However, the United States was founded on white supremacy and racial ideology. It maintained *de jure* (by law) white supremacy until the Civil Rights Act of 1964; in fact, we argue, in contradiction to remarks by Tucker Carlson on his FOX News show that white supremacy is a hoax,[22] that this ideology still plays a major role in American social life and foreign policy.

In summary, hating another group is one thing. That's ethnocentrism. Feeling that your hate is justified because people are biologically inferior (*and* science proves it) is quite another. That's racism. The idea of race is essential for racism. And the idea of race is based on deeply flawed science.

CONCLUSIONS

Race is an idea that people made real by constant use. It was a useful idea to those in power to promote slavery and colonization and, in current times, to maintain a racial status quo of differences in opportunity. Historically, scientists thought of race as equivalent to the biological subspecies concept. Today, many people still think that races are biologically different and have different attributes. However, that view of race is scientifically incorrect. Human biological variation is real, complex, and wondrous. But race neither describes nor explains human variation. Race is a social idea, not a biological one.

Our country was built on the hard work and brilliance of many. And it was also built on the myth of race that enslaved and colonized many more. In his poem "Let America Be America," Langston Hughes writes of the United States that it is "the land that never has been yet." We have never come to face the myths of race and the history of racism. We hope this chapter and chapters to follow help to bring us closer to what this land could be.

Chapter One

HOW DID RACE BECOME BIOLOGICAL?

The old, mythical idea of race as consisting of distinct biological groups of humans, what we here call *biological race* or *biologized race*, slowly took hold in Europe, then the Americas, and finally other parts of the globe. It became reified, especially by those with power and authority, eventually including scientists. Today, it seems that this dominant worldview of race as somehow rooted in biological and genetic differences is so accepted that it must be scientifically true. It seems so obvious. We can see difference. But what seems obvious can deceive.

Here, we focus on how and why this false idea of biological races developed through historical places and periods. This chapter is a short, capsulized history of the big lie: how the mythic idea that races are biological came about and became dominant. In the next chapter, we explain the science of why this idea of biologized race is incorrect and harmful.

We begin with a brief overview. It is probably true that for the vast majority of human history, each group of people thought of others as somehow different from themselves. From the first humans some 300,000 years ago to the thirteenth century, individuals rarely traveled more than forty miles from their place of birth. That's still true for many today. Individuals would rarely encounter anyone whose physical appearance was noticeably different from their own group. People in the out-group might speak a different language or dialect, have slightly different customs and ways of doing things, and rarely might even have looked a bit different. But those in the out-group were not *racially* different; that is, the differences were not thought to be innate, based on biology, and not necessarily hierarchical.[1]

Members of early agricultural groups and ancient civilizations might have considered others outside of their group as less than those in their group. Throughout history, "the other" was often mythologized as uncivilized, savage, or even cannibalistic. However, physical or phenotypic differences were of little importance. The concept of races as biologically distinct groups had not yet found a foothold.

The idea of race started around five hundred years ago, and the fully formed idea of biologized races is only about two or three hundred years old. In the next section, we outline three key steps related to the idea of biologized race: folk, legal, and scientific.

First, race developed as a *folk idea* to characterize different groups as fundamentally different types. These were general ideas used to explain similarities and differences. How these notions started is difficult to trace. They likely started with explorer tales and reports. Fantastic tales were common of wild men and bloodthirsty cannibals. The explanations these tales provided for these peoples were that they were entirely different creations (pre-Adamites) or that they were remnant, devolved humans from after the Flood.[2] In either case, the differences were fundamental and innate. That view of difference took hold in the Christian world and became common and a dominant worldview.

This folk idea of race slowly emerged over centuries and millennia. By 1492, it had become clear that a shift was in process, from seeing out-groups not as just culturally different but as something more inherent and biologically different. In that year, Jews were expelled from the Iberian Peninsula because they were not merely religiously but fundamentally different. And, of course, we all know that Columbus sailed from Spain and landed on the island of San Salvador in the Bahamas. There, and in his further explorations, Columbus and those who sailed with him encountered peoples who were remarkably different in customs, dress, language, and looks from any they had known. They must be a fundamentally different type.

Second, race became a *legal entity*. Fast-forward two hundred years after Columbus's voyages, and economies throughout the Americas became increasingly dependent on slavery and indentured servitude for labor. To stabilize these systems of oppression, race became legal. Laws in Virginia and the other colonies in the 1600s were used to legally separate Indians from enslaved Africans and enslaved Africans from indentured Europeans.

It became legal for one individual, a European, to own another, an African, based on their race alone.

Third, race became *scientific*. Guided by Christian special creationism, in the mid-1700s, naturalists made a variety of attempts to name and describe racial types. Soon after, natural scientists and early anthropologists tried to justify slavery and colonization by proving that races were fundamentally different and, by extension, that some races were meant by God to rule others. Racist science and theology made the case that races could be classified in a repeatable way and that those innate racial differences were based on biological differences in skin and hair and thought to extend to the inner body and brain. Races became inherent, universal, biological, and hierarchical.

In the nineteenth century, a handful questioned the reality and hierarchical view of race. Frederick Douglass[3] wrote that the conditions of slavery made the slave, and Charles Darwin overcame his own Eurocentric tendencies and came to the revolutionary understanding of natural selection as a process that countered idealism and the inevitability of humans' place in nature. However, it would take another century before the data would be available—and understood—to fully and scientifically dismantle racial thinking in biology and anthropology.[4]

Today, we have made progress in changing the legal status of race. Laws in the United States have eliminated racial slavery, followed by the elimination of "separate but equal" laws and the passing of voting rights legislation. Scientists have shown that biological races do not exist within our species. Yet, the folk ideology of biologized race remains. It might change from time to time and place to place. It is a chameleon of a concept. However, most fundamentally, the worldview of biologized race remains. And that is fundamentally why racism remains.

DID THE ANCIENT EGYPTIANS, GREEKS, AND ROMANS HAVE A CONCEPT OF RACE?

No.

The idea of biological races formed after these periods and places. In fact, no civilization or group from any part of the world had a concept of race at least until the late Middle Ages, from the Crusades, which began in 1095, through 1492.

The Indian caste system is older than race, and although it, too, is based on family inheritance, it is not biological in the way that race becomes biological. There simply is no record that groups were thought of as distinct because of their innate biological differences. Audrey Smedley writes, "Expansion, conquest, exploitation and enslavement have characterized much of human history of the past 5000 years or so, but none of these events before the modern era resulted in the development of ideologies or social systems based on race."[5]

"Race" or an equivalent term does not exist in the Hebrew Bible or Christian New Testament. The world of the people who wrote these texts makes it clear that they had not encountered the full scope of human biological diversity. The Levantine people, who produced the Hebrew Bible (1200–722 BCE), described persons of African descent. The prophets Nahum and Amos described Ethiopia, which meant the region south of the Blue Nile. Their description of the people of that area in Nahum 3:9 or Amos 9:7 lacked any negative aspect.

The world of the New Testament was considerably larger, which expanded in the first century with the growth of the Roman Empire. The farthest northern region in their world was Scythia (modern-day Russia), and they knew of northern Africa and Lusitania (in modern Spain) to the west. There is some evidence that Rome traded with China via the Silk Road, and some archaeological remains show that persons of Chinese descent might have inhabited the Roman Empire.[6]

Greek philosophers thought that within people could be found various ratios of humors and essences. Based on their essences, individuals were suited for specific positions in society. According to this idealist worldview, individuals and groups were essentially their specific melding of humors. That essentialist thinking provides some of the background for typological thinking. One could say this was proto-racial thinking, but it wasn't yet racial thinking, because the essences were not specific to individuals of different family groups and cultures.

The Greek Herodotus (called the "father of history," 484–425 BCE) considered non-Greeks as barbarians, but one could change their status by adopting Greek cultural norms.[7] The ancient Arabs were aware of ethnic differences but did not describe them in "racial" terms. Finally, neither Marco Polo of Venice, who traveled from Europe through Central Asia to China and back by Southeast Asia and India, or the Moroccan Ibn Battuta, who traveled through the Balkans, Black Sea, southern Russia, India, China, the Middle East, Egypt, and North Africa, described peoples they encountered in racial terms.[8] Thus, the ancients did not view others as bioracially different.

Throughout history and across the globe, it appears to have been common to use derogatory terms to name other groups and to think of one's own group as superior. This is *ethnocentrism*. Although ethnocentrism is hurtful and potentially harmful, especially as it carries over into an increasingly interconnected world, it is not racism simply because it is not based on the idea of biological races.

There is abundant evidence for this sort of thinking through the Middle Ages and the Renaissance (1300–1600). Outside groups were frequently described as barbarians and worse. In this period, European Jews were treated in ways that resemble modern racism. Yet, although there was fear and loathing of those outside one's own group, or *xenophobia*, it is also clear that humans merged and mingled through time and also evidence of acceptance, admiration, desire, love, and reproduction. Wherever humans roamed, be it in search of new lands or foods or for conquest and subjugation, humans ended up mingling and mating.

What we know of Egyptians' art and writing makes it clear that they were aware that people from different lands looked different. Egyptian hieroglyphics portray humans with a wide range of features that look like those of individuals from areas including southern Russia to the equator. They saw differences, but no concept or word equivalent to "race" existed in their language.[9] They enslaved individuals, usually the captives from wars. But there is no evidence that they thought of these differences as racial; that is, distinct biological differences. Like Moses in the Bible, one could become an Egyptian simply by living as an Egyptian.

The same is true for the ancient Greeks and Romans. Their empires expanded as they conquered foreign lands, spreading at times to the British Isles and Northern Europe and into the Middle East and North Africa. Like the Egyptians, they saw the peoples of captured lands as different in their ways of life and language but not racially different. Others might be heathens or could be enslaved, but being a heathen or slave was not based on biology or race.[10]

WHAT CHANGED IN SPAIN IN 1492?

First, the Jews and Moors were expelled. Second, Columbus left Spain and encountered the Taino of San Salvador. Indeed, the expulsion of the Jews and Moors provided some of the funds for Columbus's voyage.[11]

The Alhambra Decree, or the Edict of Expulsion, was the punctuation mark ending Jewish life in Spain, the result of over a century of religious persecution and pogroms. Issued in March 1492 by Catholic monarchs Isabella I of Castile and Ferdinand II of Aragon, the same monarchs who sent Columbus to find the Far East, the edict required all practicing Jews to leave Spain within four months. Prior to 1492, Jews had been allowed to convert to Catholicism, and an estimated half of all Spanish and Portuguese Jews, called *conversos*, did so. But the Catholic monarchs worried that *conversos* would continue to secretly practice Judaism. These Jews were called *Marranos* (swine).

These actions against Spanish Jews, as horrific as they are, were based on religious rather than racial intolerance. The expulsion might be considered to be based on proto-racial beliefs, in that a Jew will remain a Jew even if a *converso*. Jews could change their religion but not their essentialness as different. The expulsion was not based on thinking that the Jews were a race but is a sort of preview of the connection between race and racist acts.

Similarly, Columbus's contact with Taino Indians of San Salvador in 1492 is a single act. What is significant for the history of race has to do with long-distance travel, especially by boat, and exemplified by Columbus's journey. Prior to the age of European exploration, which began in the early fifteenth century, travel was typically slow and covered shorter distances by land. Travelers such as Marco Polo (1254–1324) and Ibn Battuta (1304–68/69) encountered people who were not perceivably different from people living in the lands from which they came. The differences were not profound or noteworthy. Dialects, dress, and diets changed, but they did so in ways that were subtle, continuous, and undramatic.

Marco Polo famously traveled along the Silk Road into China, discovering peoples who were different in looks, language, and customs. The same is especially true for early seafaring explorers such as Columbus and Vasco da Gama during the so-called Age of Exploration, which spanned the 1400s to the 1600s.

The effect of the sudden change from light-skinned and -eyed Europeans to brown-eyed and -skinned Native Americans almost certainly gave the impression that variation was discontinuous and that the latter had to be a distinct type.

Columbus kept a journal that included his thoughts on those he encountered in October 1492 in San Salvador. He describes the Taino by their physical features (well-made and handsome bodies), hair (like a horse's tail), and color. He described them later as simple and opined that they easily would be made Christians. He did not recognize that they had their own religion. To

Columbus, the Taino and other Native Americans were less than Europeans, to the extent that they could be enslaved. Indeed, he kidnapped and enslaved several of them to present as spoils to Queen Isabella to justify his voyage and tempted her with the lure of additional wealth that could be plundered from the new lands.[12] Although he did not yet have law and science on his side, he had an ideology that could bend to political power and that allowed him and those like him to designate *Indios,* Native Americans, as a different kind of non-Christian people, either separately created or devolved.[13]

WHAT LED TO THE IDEA OF BIOLOGICAL RACE?

Race is not the type of invention that is marked by a patent, stamped as officially invented by a person at a specific moment in history. Ideas and concepts are not invented by one person in one place and time. Rather, ideas emerge slowly and are slow to be adopted and solidified. That's almost certainly how the idea came about that different human groups are delimited by differences in biology and that those differences might also be ranked in worth. Also, the sciences of biology and anthropology did not develop all at once. As discussed by Terence Keel in his book *Divine Variations,*[14] these emerging disciplines were intimately linked with and driven by religious beliefs. It is important to recognize that prior to the nineteenth century, people simply did not have the scientific knowledge and understanding of evolutionary processes to correctly understand human biological variation.

Table 1.1 provides a timeline of important scientific discoveries and controversies that had to be resolved before the modern views of biological variation and its significance could be developed. Most important in this regard (in bold font) were the 1859 publication of Charles Darwin's *On the Origin of Species* and Gregor Mendel's discovery of particulate genetic inheritance in 1868.

Two main conceptual roots of racial thought began to come together around 1492. The first is an *idealist worldview.* Idealism was the dominant worldview up to the nineteenth century. As exemplified by Plato (427–337 BCE), idealism is a view of the material world as secondary to or derived from a more fundamental world of ideal things. Plato called these ideals forms, or *eidos.* In the history of philosophical ideas, this worldview is most often referred to as Plato's theory of forms or doctrine of ideas. In biology, it became known as *essentialism* and gave rise to the essentialist species concept (table 1.1).

TABLE 1.1
Biological Conceptions of Race Timeline

Year	Event
428–348 BCE	Plato, Greek philosopher, *eidos*—diversity of living things is the reflection of a limited number of unchanging universals.
384–322 BCE	Aristotle, Greek philosopher, scale of nature, *De Partibus animalium*.
1520 CE	Paracelsus, Swiss physician, pre-Adamite races.
1583	Andrea Cesalpino, Italian, *de Plantis*, downward classification of organisms by logical division.
1591	Giordanno Bruno, Italian priest, philosopher, pre-Adamite races.
1619	Lucilio Vanini, Italian, Africans descended from apes and had once walked on all fours.
1696	John Ray, essentialist (Plato) species concept.
1735	**Carolus Linnaeus, *Systema Naturae*, essentialist species and varieties.**
1700–1800	Naturalists disagree about African inferiority, unclear on the heritability of racial traits, effect of the environment, primarily monogenists.
1749	Georges Louis LeClerc, Compte de Buffon, *Histoire Naturelle*, forerunner of modern biological species concept, infertility criterion.
1760–1830	Lamarkian evolution, races as distinct lineages at different levels of perfection.
1800–1860	Naturalists agree on African inferiority, heritability of racial traits, polygenists.
1812	Georges Cuvier, correlation of animal parts.
1830	Paris Academy debate, Cuvier vs. Etienne Geoffroy Saint-Hilaire.[1]
1850–60	Four horsemen of craniometry: Louis Agassiz, Samuel Morton, Josiah Nott, and George R. Gliddon.
1854	Frederick Douglass, *The Claims of the Negro Ethnologically Considered*.
1859	**Charles Darwin, *On the Origin of Species, or, the Preservation of Favored Races in the Struggle for Life*.**
1868	**Gregor Mendel, principles of particulate inheritance.**
1870–73	Mitosis, meiosis, German cytogeneticists.
1871	Darwin, *The Descent of Man and Selection in Relation to Sex*, chapter 7, the races of man.
1918–51	Neo-Darwinian synthesis (unification of natural selection with Mendel's genetics); population genetics; J. B. S. Haldane, Julian Huxley's clines, Sewell Wright's F-statistics.
1930	Chromosomal theory of inheritance, Thomas Hunt Morgan and others.
1942	**Ashley Montagu, *Man's Most Dangerous Myth: The Fallacy of Race*.**
1944	DNA as material of heredity; Avery, MacLeod, McCarthy experiment.
1950–51	UNESCO race statements.
1962	Frank Livingstone, nonexistence of human races, *Current Anthropology*.
1972	**First protein electrophoretic studies of human genetic variation.[2]**
2002	Publication of human genome by Sanger sequencing technology, announced by project leaders J. Craig Ventor and Francis Collins.
2006–14	Illumina sequencing technology reduces time and cost of human genome sequencing, announces $1,000 human genome.

Note: Entries in bold represent the most significant events.

[1] The Paris Academy debate of 1830 concerned whether animals could be thought to have a unified body plan; such a plan was an idea associated with evolution and common descent. Saint-Hilaire argued for a unified body plan, opposing Cuvier, who was thought to be the most brilliant man of that age. Cuvier used the example of the human races as evidence for the separate creation of species.

[2] R. C. Lewontin, "The Apportionment of Human Diversity," in *Evolutionary Biology*, ed. T. Dobzhansky, M. K. Hecht, and W. C. Steere, 381–98 (New York: Springer, 1972).

Plato wrote that the things that were all around him, such as a chair and a tree, were not the true essences of "chair-ness" or "tree-ness." Rather, the true chair is an ideal chair, and a real tree is an ideal tree. Ideal types are an abstraction that humans must imagine, as their essences come from the gods.

There is an undeniable appeal to ideal thinking. Our brains seem to be drawn to categorizing things into their types or slots. It is not a far reach from ideal types to naming practices; for example, naming diseases provides a sense that we know or at least can recognize them. The same goes for naming types of peoples.

The first problem with idealism is that it is entirely subjective. For example, is the piece of furniture in Alan's living room a large six-legged chair, a small couch, or a loveseat? There is no objective way to decide. How many separate ideal things are there, and how does one account for intermediates? And who is to say what the ideal is? What is the ideal tree?

The second problem is that idealism does not involve a process. There is no mechanism for change: essences are static, fixed and eternal. Idealism is unscientific because the nature of things is beyond human comprehension.

Finally, idealism was a barrier to the discovery of a theory of change (or evolution) simply because few natural historians were concerning themselves with the question of change in the observable material world. Idealism was a barrier that Charles Darwin and Alfred Wallace had to overcome to work out the origin of species.[15]

That said, the idea of biological races is aided by all of those problems: idealism, subjectivity, and changelessness. It's a small step from Platonic idealism to ideal types of humans, each with a unique set of properties. The Greeks came close to thinking that there are ideal types of men and women and ideal rulers and peasants; Aristotle's concept of the "natural slave" comes to mind. Ideal types set a part of the intellectual stage for bio-racial types.

The other main root of racial thinking was later set by the "Great Chain of Being,"[16] a related view of the natural world whereby all God's creatures were seen as occupying a rung on a ladder of creation. The higher on the ladder, the closer to God. This great chain was built off Greek notions of ideals and added a god into the equation, placing animals and human groups as distinct in proximity to the deity. Some depictions of the great chain place women below men, and others clearly draw in different types of people, with Europeans occupying the higher rungs on the "Great Chain of Being." The importance of the "Great Chain of Being" is that it clearly added the dimension of hierarchy to proto-racial, idealist thinking.

DID SLAVERY MAKE RACE NECESSARY, OR DID RACE MAKE SLAVERY ACCEPTABLE?

Both!

The connection between race and slavery is central to understanding both the idea of biological races and the acceptance of *chattel slavery*, the type of slavery in which enslaved individuals are possessions of the owner. The story of the codevelopment of slavery and racial ideology is well documented in Virginia.

In August 1619, an English ship, the *White Lion*, arrived near Hampton, Virginia, and twenty or so enslaved Africans eventually found their way to Jamestown. So, slavery in the English colonies began at least four hundred years ago. In reality, racial slavery is older; countless other slave ships, mainly Portuguese, had already brought enslaved people to the Caribbean, Florida (then under Spanish rule), Brazil, and other locations in Central and South America. You can learn a great deal about the voyages that carried enslaved people to the Western Hemisphere at the Slave Voyages website.[17]

There is an important historical debate about whether these first Africans who toiled at Jamestown, as well as other early North American colonies, could be properly thought of as chattel slaves or simply indentured servants. Many Africans of this period were granted freedom after a period of servitude. The important point is that in the early days of Jamestown, it wasn't clear that there needed to be a *legal* difference between enslaved Africans and indentured Irish and other indentured Europeans.

Starting in 1661, Virginia began to pass laws that legally separated Africans from indentured Europeans[18] and made slavery a permanent state for enslaved Africans. This was done to reduce the organizing power of both groups. Although indentured Europeans did not have political power, at least they eventually could become free. This was the first time European or Christian was established as an identity and in opposition to the label of Native American or enslaved African.

Is race the parent of racism, or is racism, in the form of slavery and colonization, the parent of race? Both. The idea of race was made out of political necessity to justify continued enslavement and colonization.

HOW COULD EUROPEAN COLONISTS JUSTIFY ENSLAVEMENT AND THE DECIMATION OF NATIVE PEOPLES?

Race.

The concept of race justified the inhumane treatment of people of different races. At the time Columbus arrived in the Caribbean, European society was emerging from feudalism. In this social system, the members of the ruling class had relatively absolute power over peasants and artisans. Europeans felt that they had the right to conquer non-Christians, and with the voyages of discovery, more and more non-Christians who were not white were being discovered. Originally, the debate was whether these new people had souls and could be converted to the Christian faith.[19] The question of the status of non-Christians has its roots in debates from the Middle Ages and the possibility of redemption for Jews and Muslims in Spain. The forced conversion view had its roots with Aristotle and his idea of the natural slave and St. Augustine's intolerance of other religions.

The opposition to forced conversion and conquest of the American Indians came mainly from the Dominican order of the Catholic Church. Dominicans opposed such conversion on the basis of both natural and divine law. The opposition included Francisco de Vitoria, considered the father of international law; Bartolomé de las Casas, bishop of Chiapas and apostle of the Indians; Domingo de Soto, a theologian at Salamanca University; Bartolomé Carranza, bishop of Toledo; and Antonio de Montesinos, a friar on the Island of Hispaniola whose preaching led to the conversion of de Las Casas. The Dominican appeals to the Catholic kings led to the Laws of Burgos of 1512. That law proposed that:

(a) the American Indians were free,
(b) they were to be instructed in the Christian faith, and
(c) the Spanish monarch could utilize them to work, but the work must be of a nature that would not be detrimental to their physical or spiritual well-being.

However, secular interests overpowered the spiritual intentions of the church, and the laws were never implemented. Thus, the debate during the seventeenth into the nineteenth century was not about whether the Indians had souls, but how to best justify their being decimated and enslaved. This discussion was closely related to biblical questions on the origin of human races.

Essentially, there were two answers to the question of how the races came to be: *monogenism* and *polygenism*. Monogenists believed that all humans were created by God at the same time. They followed the Genesis story in the Bible: we are all descendants of Adam and Eve. However, the races separated after the sons of Noah dispersed after the great Flood. The different races were descended from different sons; Africans were descended from Ham's son Canaan, who was cursed by Noah.

It is interesting in this context to understand that Genesis does not mention which sons gave rise to which races. This is easy to understand because, as we discussed earlier, the ancients had no concept of race. The tribes mentioned in Genesis all inhabited the Levantine world, and the descendants of each son overlapped in their residence within the Mediterranean basin, Middle East, and Arabian Peninsula. This explains why medieval scholars consistently disagreed about which people were descended from which son. The idea that the curse of Ham was visited solely onto the "darker races" took hold after the age of discovery and was used as an argument to justify the enslavement of Africans.[20]

In polygenism, the different human races were created separately (the pre-Adamite races; see table 1.1). The Bible is not necessarily incorrect, it just isn't about those other creations.

In monogenism, the races are closer to different varieties of humans, and in polygenism, they are closer to the modern notion of different species. In some sense, polygenism offered the strongest ideological cover for subjugation, because it suggests that the different races are outside of the Bible and more like different species. In this scheme, killing an individual of a different race would not be murder but equivalent to killing a hog or dog.

In summary, both monogenists and polygenists owned slaves and colonized lands. Monogenists might have made more of an effort to rehabilitate individuals of different races and turn them into Christians. But everyone bought into the justification for inhumanity. In both monogenism and polygenism, races were considered separate and unequal. The difference was in degree: greater in polygenism. For everyone, the idea of races as separate and unequal had high stakes. It still does.

HOW DID EIGHTEENTH- AND NINETEENTH-CENTURY SCIENTISTS TURN A FOLK IDEA OF RACE INTO A SCIENTIFIC CONCEPT? WHY DID THESE SCIENTISTS DO THAT?

Scientists didn't invent the idea of biological races among humans. But they certainly did a lot to solidify the reification of race, to make race seem unquestionably real. Before scientists got involved, race was a folk idea, and then it became a legal entity. A folk idea is one that is common but does not necessarily come from any one place; it is just widely shared, common wisdom. That was likely part of what made race seem real. It was simply that different peoples are essentially different.

Then race became a legal entity. It became legal to separate race in the colonies. Being white eventually gave poor whites legal rights and social status—to vote, own land, own other people, and a lot more—that were legally denied to individuals of other races.

And then scientists got into it! Scientists were instrumental in moving race from a folk idea and legal entity to an unquestioned real thing. However, it must be remembered that biology and other sciences in this period were practiced by individuals who believed in the literalness of the Bible and were members of the Christian church. The schism between Christian and biological views of human origins and diversity did not begin until well after the publication of *On the Origin of Species* in 1859.

Before science got into the race business, races were already seen and accepted as discrete and unchanging types with separate places on the Great Chain of Being. That view had been made real by constant use. The power of rulers and the clergy, state and church, made it true. Virginia laws then codified it and made it legally true.

But some still wondered if it was natural fact that races would be forever separate and unequal. At the end of his *Voyage of the Beagle* (1839), a young Charles Darwin asked, "If the misery of the poor be caused, not by the laws of nature, but by our institutions, [then] great is our sin."[21] Science was employed by those in power who asserted that it was indeed nature that made, and would continue to make, misery.

It is important for the reader to understand that what we think of as science, and specifically biological science, in the eighteenth century is vastly different from modern science. In the Western world, biological and anthropological science at that period was not independent from theology. Thus, the origin of human beings was thought to be the result of a divine act of special creation,[22] and explanations for the different biological characteristics of human beings had to relate to the book of Genesis. As noted above, those naturalists who felt that humans of all types shared common descent were called monogenists. Thus, in this worldview, the differences observed in human beings had to do with degeneration resulting from either the mark of Cain (Gen. 4:15) or the curse of Ham (Gen. 9:25–27). Samuel Stanhope Smith, a Presbyterian minister and professor of moral philosophy at the

College of New Jersey, wrote in 1788 that all humans were descended from Adam and Eve and that differences in human complexion and form were the result of climate, geography, and custom.[23]

Mary Wollstonecraft, best known for *A Vindication of the Rights of Woman*, was also a monogenist.[24] Most of the prominent naturalists of the eighteenth century were monogenists, including Carolus Linnaeus (1707–78); Georges Louis LeClerc, Compte de Buffon (1707–88); and Johann Friedrich Blumenbach (1752–1840).[25]

Polygenist (or the pre-Adamite race concept) ideas began to surface in the sixteenth century. Isaac La Peyrère wrote *Pre-Adamite (Men before Adam)* in 1655. By the nineteenth century, the tenor of naturalist thought had shifted to favor this worldview.[26] As this view was considered in opposition to the Christian scriptures, it caused considerable controversy.[27] However, prominent men of science such as Louis Agassiz and the Episcopalian Samuel George Morton (1799–1851) backed it.[28] This is one of the reasons that Darwin did not discuss human beings in *On the Origin of Species*.[29] Not until 1871, when he had achieved the position as one of the most prominent scientists, did he dismantle the pseudoscience of polygenism in *The Descent of Man*.[30]

The science of taxonomy (classification of living things) goes at least as far back as the ancient Greeks. The main principle used in this science by the sixteenth century was the idea of logical division. This simply means that if two organisms share a common characteristic, they are included in the same group. This, of course, requires the scientist to decide which traits are important to classification. Note that classifying organisms together does not require that one think the organisms share a common descent. The early taxonomists thought that the groups were fixed, eternal, and unchanging and based on the Great Chain of Being. For example, when Italian anatomist Andrea Cesalpino started classifying plants (see table 1.1), he did so within a creationist framework. Similarly, European naturalists such as John Ray thought they were cataloging God's creation. Linnaeus began his work in the 1730s, and his purpose was to develop a system to classify all organisms (not just humans). Similarly, leading French naturalists such as Comte de Buffon, Georges Cuvier, and Jean-Baptiste Lamarck were not simply interested in the classification of humans; rather, they were attempting to develop a comprehensive system of all living things within the framework of natural theology.[31]

Thus, when taxonomy was applied to humans, it also was driven by ideas that had been developed and applied to all of the natural world. Of course,

European naturalists also brought into the process biases that had been developed due to manifestly unequal social conditions. So, when the science of taxonomy was applied to humans, it had two main goals. The first goal was to figure out the names, numbers, and essential characteristics of human varieties (or races). Once those were classified, some eighteenth-century scientists had a second goal, which was to prove by the rather newly adopted methods of science—measuring and comparing things—that the races were unequal and had always been so.

It is important to note that not all eighteenth-century European naturalists felt that human races were ranked hierarchically, François Bernier of France (1625–88); Gottfried Wilhelm von Leibnitz of Germany, co-inventor of calculus; and Blumenbach of Germany considered the father of anthropology, did not think that human races should be ranked hierarchically.[32] Leibnitz's personal experience convinced him that Africans were not inferior to Europeans. One of his most gifted students was Wilhelm Anton Amo. Amo was born in Ghana and raised as a house boy (read as "pet") of a German noble. He went on to study mathematics at university. Amo's career never developed, as there was no place in eighteenth-century European society for an African mathematician. He eventually died penniless on the west coast of Africa.[33]

Many scientists named, numbered, and described the races. This was a cottage industry in the seventeenth to the nineteenth century. These early scientists were known as *naturalists* for their study of the natural world, including the natural varieties of humans. The aforementioned Bernier, a physician and amateur naturalist, might well have been the first to develop a clear and comprehensive classification of human races. His races are more akin to species: Europeans, Far Easterners, Negroes, and Lapps. Yes, Lapps. Bernier obviously was not very familiar with much of the world.

Until the 1800s, it was clear to the Christian world that God must have created some number of different human types/species (polygenism) or that some human types degraded after the Flood (monogenism). Remember, this is a world in which everything seemed to be fixed and more or less as God created. Thus, no one gave much consideration to the problem of intermediates. For example, the idea at that time of gradual variation that we know so well today made less conceptual sense than seeing the world as consisting of distinct types. The goal was not to understand the process by which things came about; that was a given: God created the world. The classifiers' goal was to understand which things were created.

One of those who wondered about intermediate forms was German scientist Blumenbach the "great-grandfather" of anthropology. Blumenbach organized humans into five races: Caucasian, Mongolian, Malayan, Ethiopian, and American. He is credited with naming the white or European race "Caucasian" after a beautiful skull from the Caucasus mountains. Although Blumenbach's concern for intermediates was shoved aside, his classification of five races was influential, and the name Caucasian stuck.[34]

Linnaeus was a Swedish naturalist who is credited with the notion that races are separate subspecies. Linnaeus traveled in his early life and later relied on reports from explorers and conquerors. He is the father of modern taxonomy and envisioned himself as having categorized what God created. God created, and Linnaeus classified.

Linnaeus's race categories included *Americanus, Africanus,* and even *Monstrosus* (for wild and feral individuals and those with birth defects). To Linnaeus, their essential defining traits included a biocultural mélange of color, personality, and modes of governance. Linnaeus described *Europeaus* as white, sanguine, and governed by law and *Asiaticus* as yellow, melancholic, and ruled by opinion. These descriptions highlight just how much ideas of race are formulated by social ideas of the time. The goal was to describe and classify. Many did.

Around 1800, the center of race science shifted from Europe to the Americas and from classification and naming the different racial types to describing the essential, everlasting hierarchical differences. The shift happened for a simple reason: the need to justify slavery and colonization and, after slavery ended, institutional and structural inequalities.

In the United States, the contradiction of the country's freedom from England and its enslavement of Africans was all too apparent. The Abolitionist movement gained steam at about the same time as the war for independence from England. Abolitionism was a movement of importance, with ethics on its side. How could one man ethically enslave another? Perhaps they could do so if the other was meant by God to be controlled.

Here, the work of Samuel George Morton is illustrative and central. It also demonstrates the growing ideological power of science to make something real. Morton, a Philadelphia physician and self-styled natural scientist, was famous for collecting skulls from around the world. He was not seen as a thorough racist scientist. Rather, working in Philadelphia, he was considered to be neutral, objective, and data driven regarding questions about race and intelligence.

Thinking that intelligence is the most important human capacity and that it could be measured by the size of the brain, Morton set about measuring the brain sizes—or, more precisely, the cranial capacities—of skulls that he had collected of different races, including ancient and contemporary Egyptians and Native Americans. He wanted to show that (a) the races could be separated and ranked on cranial capacity (= intelligence) and (b) those rankings were permanent through time.

Morton tabulated the cranial capacities of different "races," both in the present or near term as well as in the past, by measuring ancient Egyptians and Native Americans. Showing also that ancient skulls had cranial capacities comparable to their contemporaries made it clear to Morton that his results supported polygenism. Morton's work, emerging just before the Civil War, was influential. One of the only individuals who stepped up to challenge Morton was Frederick Douglass.[35] Douglass, born a slave, taught himself to read and had very little formal education. Yet, he brilliantly saw that one's conditions in life helped to create the person one became. Environments worked to shape man.

Douglass's writings point clearly to the importance of environment in shaping organisms and in this sense is early evolutionary thinking. Moreover, he was also aware of the biases in writing: "Scientific writers, not less than others, write to please, as well as to instruct," "and even unconsciously to themselves (sometimes) sacrifice what is true to what is popular." Finally, although we focus on Morton, it is important to know he was not alone. Other individuals were to echo his results and, more so, his sentiments.[36]

Picking up on Douglass's insights into bias and before we move on, you might be aware that Morton's results were re-examined by Steven Jay Gould (1941–2002).[37] Gould claimed that Morton finagled his results to fit his preconceived notions. Others think that Gould might have also finagled his restudy of Morton. What's important is not Morton's objectivity. After all, his equating of cranial capacity with intelligence is incorrect. Rather, what is important is that science was used to prove what the culture of those scientists supported: slavery and polygenism.

DID THE END OF SLAVERY CHANGE IDEAS ABOUT RACE?

No.

In fact, what changed was that the end of legal slavery increased the fervor of those in power to prove the races are separate and unequal and thereby to justify other forms of racism. Races became more equal under the law, but those laws were often averted. Races remained unequal in power and all aspects of life. With less legal support for slavery, scientists worked harder to prove that races were inherently unequal. Indeed, Agassiz, one of the most virulent racists in the history of American science, was hired to work with the Freedman's Bureau after the Civil War had concluded.[38]

The ideology of race certainly did not end when slavery was declared to be illegal with the passage of the Thirteenth Amendment of 1865. Moreover, no longer protected by the legal sanctioning of slavery, many scientists continued to study racial differences with preconceived notions of racial hierarchies.

There are many examples of the continuation of racist science. Efforts were made to find the root of racial differences in all sorts of glandular substances and parts of the body. Researchers continued to focus on efforts to prove differences in intellectual abilities both by examining brains and testing cognitive abilities.

The research of Robert Bennett Bean (1974–44) provides a clear example of the shift from skulls to brains and the continued drumbeat of racist science. Bean, a professor of anatomy at the University of Virginia, wrote extensively on subjects such as the biological uniqueness of the Philippines, the Jewish nose, and black brains. In 1906, Bean published an eighty-page article titled "Some Racial Peculiarities of the Negro Brain" in the *American Journal of Anatomy*. This is a piece of science that has all the trappings of objective science: there are lots of numbers, and its conclusions are clear and precise. Bean makes four concluding points:

1. The brain of the American Negro is smaller than that of the American Caucasian, the difference being primarily in the frontal lobe, and it follows that the anterior association center is relatively and absolutely smaller.
2. The Negro brain can be distinguished from the Caucasian with a varying degree of accuracy according to the amount of admixture of white blood.
3. The area of the cross section of the corpus callosum varies with the brain weight. However, in the Negro its anterior half is relatively smaller than in the Caucasian, to correspond with the smaller anterior association center; the genu is relatively larger and the splenium relatively smaller.
4. From the deduced difference between the functions of the anterior and posterior association centers and from the known characteristics of the two races the conclusion is that the Negro is more objective and the Caucasian more subjective.

The Negro has the lower mental faculties (smell, sight, handcraftsmanship, body-sense, melody) well developed, the Caucasian the higher (self-control, will power, ethical and Esthetic senses and reason).[39]

Bean's career, including his service as chair of the Department of Anatomy, is honored by the Robert Bennett Bean Award of the University of Virginia School of Medicine. The award has been given annually since 1968 to a member of the faculty for excellence in teaching the basic medical sciences. Bean is buried in the university cemetery.

Work such as Bean's in the tradition of anatomy and physical anthropology was soon to be joined by work in the growing field of psychology and mental testing by paper-and-pencil intelligence tests, and ultimately by the growing field of hereditarian studies and eugenics.[40]

IN THE TWENTIETH CENTURY, WERE JEWS, ITALIANS, AND IRISH THOUGHT TO BE SEPARATE RACES?

Yes, but not in any consistent way.

Efforts were made to characterize different groups of Europeans as different races. For example, William Z. Ripley characterized some dozen races of Europeans in his "The Races of Europe" (1899),[41] and Carleton Coon did the same forty years later in his 1939 book of the same title.[42] In between these books were efforts to restrict immigration from Southern and Eastern Europe and elevate the Nordic and Aryan racial type, and the genocide of Gypsies and Jews as degenerate races.

Grasping at straws, practitioners of racist science tried to restrict the composition of the club of whiteness and guard the borders between white and nonwhite. The rise of eugenics in the 1910s and 1920s led to efforts to protect the purity of the white race, white being so-called Native Born whites. This purity was to be maintained by restricting immigration from Eastern and Southern Europe and passing laws to prohibit marriage between individuals of different races. Scientists such as Bean were explicit and unequivocal in their desire to prove the superiority of the white race. But who is white? And is there also a ranking and separation of European racial types?

The greatest efforts were made to prove the separateness of Jews. Bean pitched in and wrote on the unique features of the Jewish nose: "The nose of the Jew is large. . . . That it is a Jewish feature cannot be doubted."[43]

In 1927, E. O. Manoiloff of Leningrad published "Discernment of Human Races by Blood: Particularly of Russians from Jews" in the *American Journal of Physical Anthropology*.[44] Manoiloff claimed that Jews and Russian blood can be identified by adding reagents and watching their color change. He proposed this method because it was difficult to make this discernment based on physical features. Although Manoiloff wrote that he was successful, his methods were entirely nonsensical and easily disproven. The paper is a fraud and should have been so recognized by the journal editors. This is the type of science that led Hitler to believe that the Jews were a separate race that needed to be expunged from the Aryan population.

After the 1960s, eugenics began to fade, and race once again solidified along the lines of protecting whites. Although prejudice still exists against "Whites of a Different Color"[45] in the United States, the category of whiteness expanded to include the Irish, Polish, Jews, and others from Southern and Eastern Europe.

ALTHOUGH CULTURAL IDEAS ABOUT THE NUMBER AND TYPES OF RACES CHANGE, WHY DOES THE FUNDAMENTAL CONCEPT—THAT HUMANS ARE DIVISIBLE BY DISCRETE BIOLOGICAL GROUPS—PERSIST?

Scientists were never able to establish a replicable classification of races because human variation is simply not racial. In the absence of an objective method, cultures are free to make up racial classifications to fit their needs.

What needs to be understood is racial classifications as a cultural category are subjective and will change over time and place as the dynamics of racial interactions and racism change. These changes are seen in different classifications of race in different countries and even in the change in official classifications over time in the United States.

Yet, most Americans think that racial classifications are based in biology. Biological race persists because it is a powerful ideology for maintaining the status quo. If, for example, one can explain racial differences in infant mortality as due to genes and inherent differences, rather than lived experiences and racism, then one need not address this inequality. It is natural. A lot is at stake in the belief in inherent biological differences, and that is why it is so intractable.

CONCLUSIONS

The existence of biological races in our species came to be the dominant worldview. This worldview started to gain traction around 1492 and did so in three phases: folk ideology, backed by the idea of types and a racial hierarchy; legal separation of races; and scientific reification. Fully formulated biological races (subspecies) did not come about until the 1800s and was most virulently promulgated by pro-slavery scientists and those in support of racist Jim Crow and anti-immigration laws. Today, the idea of biological races has been disproved by science. Yet, the worldview that race is based on biological differences persists and continues to hold back social justice and science.

Chapter Two

EVERYTHING YOU WANTED TO KNOW ABOUT GENETICS AND RACE

This chapter focuses on the science of human genetic variation and why humans are not divided into biological races. The big truth is that race neither explains nor describes human genetic variation. Race-as-biology is a myth and should never be used as a substitute for human genetic variation.

To understand why humans do not have biological races, we clarify the evolutionary forces that create and maintain genetic diversity (mutation, natural selection, and genetic drift). In case you're concerned about getting bogged down in a lot of scientific lingo, we will attempt to explain these concepts using clear, accessible language. We also encourage you to look up any terms you don't understand in one of the online glossaries that we referenced in the preface.

We then point out why common notions of race are completely at odds with the modern scientific understanding of human variation. Along the way, you'll be introduced to important concepts such as continuous variation and discordant variation. We explain the importance of understanding the structure of human variation and how it is measured. Finally, we state the reasons that the existence of regional biological variation within our species is not equivalent to the existence of biological races within our species. First, let's start with some basic genetics!

WHAT IS THE HUMAN GENOME, AND HOW DID
WE COME TO READ IT?

To understand differences among individuals and groups, let's make clear what is being measured: differences in human genomes. A genome is an organism's complete set of DNA. The complete human genome contains about 3.3 billion base pairs of DNA strung together on twenty-three pairs of chromosomes of different sizes. One set of chromosomes is inherited from the biological mother and the other from the biological father, totaling 46 chromosomes. The bases are paired sugars that make up the structure of DNA. The four base pairs that you might be familiar with are adenine (A), cytosine (C), guanine (G), and thymine (T).

Everyone except identical twins varies genetically.[1] Recent advances in human genetics have shown us just how similar all humans are genetically.[2] On average, about 999 of the 1,000 letters in base pairs of the genome are the same among individuals. Alan and Joe, or Alan and you, or Joe and you, whoever you might be, are about 99.9 percent similar genetically. This means that each person differs from any other in about 3.3 million base pairs. So, two questions arise. First, are two individuals who are purported to be in the same race significantly more alike than two individuals in different races? Second, what is significant?

Knowing that letters are different is akin to having a list of individuals who attended a party. Useful information that might be, but if you want to solve a murder that occurred at the party, it is important to know what every individual was doing at the time the murder was committed. Gene expression is like knowing what every individual was doing at any given time at the party. In organisms like us, our DNA molecules are associated with a group of proteins called *histones*. Histones play an important role in how these genes are expressed in various tissues throughout the body. Of the 3.3 billion base pairs in your DNA, only about 1.5 percent of them are directly involved in coding for a protein. Another 7 to 8 percent regulates the expression of the protein-coding DNA; in other words, about 90 percent of your DNA is *not* related to encoding your proteins. Going back to our murder analogy, the 90 percent that are non-coding could not have been involved in the crime. However, this noncoding DNA plays an important role in how *genetic ancestry*, the genetic connections that have developed through history, is determined by genetic tests.[3]

The second question is a tricky one, in part because "significant" has multiple meanings, such as in statistics and significance with regard to a process or context. For example, is being 99.92 percent genetically similar significantly different from being 99.88 percent similar? The short answer is, it depends. So, consider this a tease, and let's save any more of this discussion for later.

It is amazing that we can answer the first question about how similar each of us is genetically, thanks to a lot of hard work, technical breakthroughs, and creativity. It is answerable because we have now read the full genomes of tens of thousands of individuals from around the globe.

Here's some backstory to how we got here. Do you remember The Human Genome Project (HGP), which started in 1990 and ended in 2003? As a reminder, it was the largest collaborative biological project in history, with a final tally of $3 billion. The first draft of the human genome was announced as complete on June 26, 2000, making local and national news. It was on the front page, above the fold, of the *New York Times*. The accomplishment was vaunted as unraveling the long book of life. That was hyperbole; it really isn't the book of life, but it was important. To our surprise, Craig Venter, president of Celera Genomics and one of the captains of the HGP ship, announced on that day that the new genomic information showed that race was a myth.

Actually, we knew already that race was a myth based on prior studies of blood groups and the like, which we discuss later in this chapter. And his finding, based on three individuals, added little to the scientific proof that humans do not have biological races. But it was a headline that we agreed with, and it did usher in many subsequent studies that examined the DNA of many more individuals from different groups of peoples. These studies put an exclamation point on the short declarative statement that "race is a (biological) myth."

After 2003, the National Institutes of Health's National Human Genome Research Institute (NHGRI) initiated a program to bring down the cost of whole genome sequencing to $1,000 in ten years. This accelerated the development of new sequencing technologies (called *next generation sequencing*, or NGS), resulting in unimaginably fast genome sequencing speeds. By 2014, the $1,000 genome had been achieved. Since the dramatic reduction of sequencing costs, hundreds of studies of human genetic variation have been conducted. These studies compared the nucleotide sequences of specific genes, as well as entire genomes, including comparisons of the genomes of populations from all over the globe. The results further validated Venter's original claim that the notion of biological races within our species is a myth.

HOW MUCH DO HUMANS VARY GENETICALLY FROM ONE ANOTHER AND BY POPULATION?

The ultimate sources of all genetic variation are mutations. *Mutation* refers to random changes in the genetic code due to an error in DNA replication. The enzymes responsible for replicating DNA are like any other machine: they usually replicate DNA with great fidelity, but every now and then, mistakes happen. The mutation rate for organisms most like ourselves (mammals) is incredibly low, about 1 mutation per one hundred million (10^8) base pairs.

From that point, the fate of mutations varies. The vast majority of mutations within coding regions of genes are harmful and are therefore removed or kept at low frequencies by natural selection. Next in frequency are neutral mutations; many of these are found in the noncoding portions of the genome. This is where most genetic variation will be observed in any species. This variation is largely neutral, but it is useful for determining ancestry (see chapter 9). Finally, a small proportion of mutations are beneficial, and these are increased in frequency by natural selection. Their frequency will differ by geographical region as well as other aspects of environments (such as diet or the presence of parasites).

As a result of the NGS revolution, we know that on average about 99.9 percent of the genomes of all living humans are the same.[4] There may be a higher amount of difference between individuals due to copy number variants (CNV), making humans 99.4% similar in these variants. Copy number variants result from the genetic processes of duplication, deletions, insertions, and inversions.[5] The global differentiation of SNPs and CNVs are pretty much the same, meaning that there is more variation within populations than there is between populations for these types of genetic variants. Of those differences, the majority are shared in all populations across the globe, smaller amounts are unique to continental origin, such as Africa, Asia, or Europe; and even less are unique to a region within a continent, such as East or West Africa.[6] The distribution of these differ by population.

HOW DO SCIENTISTS EVALUATE GENETIC DIFFERENCES AMONG GROUPS OF INDIVIDUALS?

There are a number of ways to think about, and tools to measure, the genetic similarity (or difference) among groups (or populations). One of the most tried and best known is the *population subdivision statistic* (F_{ST}), which was invented by the American population geneticist Sewall Wright (1889–1988).[7] This statistic quantifies the difference in the frequencies of alleles in a population compared with the allele frequency of the entire species. Remember that at the level of the DNA, an allele is an alternative spelling of a gene. For example, sickle cell allele is an alternative spelling of the gene that codes for the beta chain of the hemoglobin molecule.

F_{ST} is mathematically simple. F_{ST} is the variation within a population over the total variation. Results can range from 0 (no population subdivision) to 1.00 (complete population subdivision). The higher the F_{ST}, the more differences among populations. For a number of theoretical reasons, Wright felt that an F_{ST} value exceeding 0.25 would indicate the presence of something akin to subspecies or biological races within a species.[8]

Whole genome studies of humans show that our population subdivision between major continental groups is much less than 0.25. Exactly how much lower varies by study and what is being studied. For example, it is about 0.110 for protein-coding genes.[9] These values are much smaller than those observed in other large-bodied mammals.

However, it would be a mistake to think that the F_{ST} values for all human genes are low. Humans display geographically based genetic variation, such that some genes are strongly differentiated between population groups. For example, for one single nucleotide polymorphism (SNP) associated with a greater risk of Crohn's disease (rs10761659), the global F_{ST} is 0.351, with the frequency of this risk allele worldwide at 0.542. In Africa it is 0.015; the Middle East, 0.427; Europe, 0.507; and East Asia, 0.759. For another SNP associated with Crohn's (rs100777885), the F_{ST} is 0.062, with the frequency of the risk allele worldwide at 0.820; in Africa, 0.975; the Middle East, 0.809; Europe, 0.812; and East Asia, 0.898.[10] In the former case, the frequency in Africa is different from the rest of the world, whereas in the latter case, the frequencies are similar in these different regions of the world.

HOW DOES THE TOTAL AND APPORTIONED HUMAN VARIATION COMPARE WITH THE VARIATION AMONG GORILLAS AND OTHER MAMMALS?

We humans are a young species. One clear result is that we have very little genetic variation compared with that of other species.[11] What's most interesting about humans is that despite their global range, they exhibit less variation

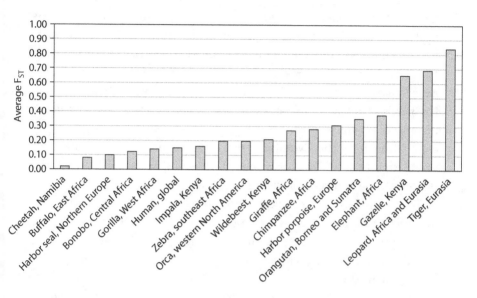

FIGURE 2.1. Average F_{ST} values for humans and other large mammal populations. *Source*: Courtesy of Alan Templeton.

among regions and continents than most other mammals. For example, the human F_{ST} of about 0.15 is considerably less than in geographically limited species such as orangutans and gazelles (fig. 2.1).

The chimpanzees of tropical Africa also contain within-species variation. These animals are considered to have subspecies or geographical races. The F_{ST} value between Western and Nigerian-Cameroonian chimpanzees is 0.270, greatly exceeding Wright's 0.250 threshold.[12]

The amount of genetic variation within any species is determined by the mutation rate, the amount of time the species has existed, and demographic events such as a population bottleneck that occurred during the time that it existed. There is some evidence that mammalian mutation rates differ by climate, such that mammals that evolved in the tropics have slightly higher rates than those that evolved in the temperate and Arctic zones. Thus, we would expect humans (who evolved in the tropics) to have slightly higher rates than polar bears (which evolved in the Arctic).[13]

Modern humans have been on this planet for about 300,000 years. This duration is typical of animal species. About 90 percent of all existing species formed in the last 100,000–200,000 years.[14] This is consistent with what we know about the fossil record of complex organisms on Earth. It shows that most complex species—especially hominids—had relatively short durations in the history of life. Examples include *Homo ergaster* (1.8–1.5 Mya, duration 300,000 years), *Homo erectus* (1.2–0.4 Mya, duration 800,000 years), *Homo heidelbergensis* (0.6–0.2 Mya, duration 400,000 years), and *Homo neanderthalensis* (0.3–0.03 Mya, duration 270,000 years).[15] Thus, the small (0.1 percent) difference between any two humans who currently reside on this planet is similar to what we observe in most other animal species. By comparison, if genetic diversity is simply a function of species duration, we would expect that *H. erectus* was probably the most diverse and *H. neanderthalensis* the least diverse of all human species.

Thus, considering that 200,000 of our 300,000-year existence was spent in Africa, it makes sense that sub-Saharan Africans contain more genetic diversity than all other human groups. In addition, non-African humans left Africa in the late Pleistocene and lost genetic variation due to the founder effect (genetic drift).

Comparing the sister species of chimpanzees (*Pan troglodytes*) and bonobos (*Pan paniscus*) helps to understand how the demographic history of a species affects its genetic variability. These species share a recent common ancestor with humans, between five and thirteen million years ago. The two chimp species are thought to have diverged approximately one to two million years ago and have experienced different demographic histories since their divergence. The history of bonobos is marked by population bottlenecks (sharp reduction in numbers) in the past and a smaller long-term effective population size. They were also geographically isolated in central Africa for a long time. Bottlenecks, isolation, and small population size all serve to decrease genetic diversity.

On the other hand, chimpanzees inhabited a much wider region across Africa (from Tanzania to Guinea, at the east and west edges of Africa). Most chimpanzee subspecies, except the western population, had larger population sizes, and not surprisingly, they maintained a greater level of genetic diversity compared with bonobos. The western chimpanzees are thought to have spread from a very small ancestral population, whereas central chimpanzees continuously inhabited central Africa. Thus, these demographic events mean that chimpanzees are more genetically diverse than their sister species, bonobos.[16]

WHAT EXPLAINS HUMAN GENETIC DIFFERENCES?

As we discussed previously, genetic differences in our and all other species start with mutations. A combination of selective and random events then determines the frequency of the mutations in any given population. Our species originated in sub-Saharan Africa and stayed there for 200,000 of its 300,000-year existence. Due to climate changes in the late Pleistocene, humans were able to migrate out of Africa. Migration also occurred within Africa. It is the second-largest continent and spans the greatest amount of latitude compared with all other land masses (37° N to 35° S). Today, there are at least seven vegetation zones: tropical forest, wooded savannah, subtropical moist lands, subtropical dry lands, thorn brush vegetation, low altitude desert, and highlands.

At the end of the Pleistocene, the Sahara occupied considerably less area than its current confines.[17] This means that human populations that migrated within Africa faced new environmental conditions. This in part explains why sub-Saharan Africans display greater genetic diversity than all other world populations. More time equates to more opportunity to build up variation.

The migration out of Africa was also bidirectional. Studies of ancient DNA indicate that back-and-forth migration occurred between regions within sub-Saharan Africa, between sub-Saharan Africa and the Middle East, and between the Caucasus/Southern European regions and sub-Saharan Africa during the first great migration of modern humans.[18] The migrating populations were smaller than their originating populations. Thus, by statistical necessity, these migrants contained less genetic variation (founder effect/genetic drift) than the populations from which they originated.

Migration between populations is an additional factor that works against strong genetic differentiation. Thus, gene flow among human population centers is one of the reasons that humans never formed geographically based biological races (or subspecies). Unlike other species that separate geographically and, once separated, remain so, humans have a history of back-and-forth migration and continual interchange across geographic areas.[19]

As human populations established themselves in new regions, they also began to accumulate new mutations. Some of these mutations were beneficial and increased until fixation in these populations. *Fixation* means that all individuals within the population would have the same allele at the locus in question. As all new mutations are rare, the rate at which they achieve fixation is dependent on their evolutionary fitness or differential survivorship and reproduction (see fig. 2.2). Generally, in nature, the force of selection on complex traits is very weak. Thus, changes in traits like behavior take a very long time. Genomic studies over the last decade have allowed us to gain a better understanding of how much and where adaptation occurred in our species over the last 200,000 years.[20] A list of these traits appears in table 2.1.

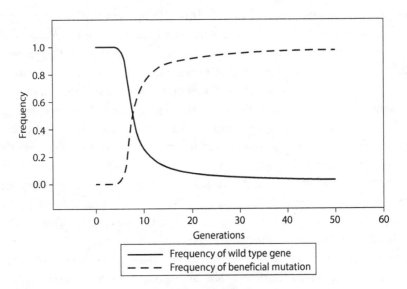

FIGURE 2.2. A selective sweep of a rare beneficial mutation with a strong fitness effect. As the beneficial mutation increases in frequency, the "wild type" gene must decrease as the frequency of $p + q = 1$, where p is the frequency of the wild type and q is the frequency of the beneficial mutation. The algebra governing this relationship shows that even a weakly beneficial mutation will eventually replace the wild type. The number of generations required for this to happen is a function of the selective pressure.

Most of them are not surprising. Humans adapted to altitude, new climates, diets, pathogens, solar intensity, and toxins. What might be surprising to some is that these adaptations generally have a relatively simple genetic basis. For example, a single mutation, a substitution of A to T at the sixth position on the beta chain of the hemoglobin molecule, causes an amino acid change from glutamic acid to valine. That change leads to a different folding of the molecule that confers resistance to malaria. These types of mutations are rare but do occasionally occur!

Skin color variation is greatest among Africans, considering that Africa passes through the largest latitude. One variant associated with fair skin, SNP rs1426654 (G to A), is found in modest frequency in East Africa (0.571 in Somalis), but it went through strong selection in fair-skinned populations such as Europeans, Central Asians, and North Indians (frequency of A = 1.00).[21] Thus, modern human genetic diversity is explained by natural selection for variants, allowing adaptation to various features of the environment in different portions of the globe, combined with aspects of population history, as further explained in the next question.

What explains variation? In the end, it is the processes of adaptation to local conditions, gene flow, and genetic drift, all of which are dependent on

TABLE 2.1
Adaptation in Modern Humans

Trait	Genes	Locus	Locations
Lactose tolerance	*MCM6*	2q21.3	E. Africa, ME, N. Europe
Arctic environment	*CPT1A, LRP5, THADA, PRKG1*	11q13.3, 11q13.2, 2p21, 10q11.23-10q21.1	Siberia
High-fat diet	*FADS1, FADS2, FADS3*	11q12–11q13.1	Greenland
Thick hair	*EDAR1*	2q13	E. Asia
Starchy food	*AMY1*	1p21.1	E. Asia
Skin pigmentation	*SLC24A5, SLC45A2, TYR, MC1R*	15q21.1, 5p13.2, 11q14.3, 16q24.3	Europe
High altitude	*VAV3, ARNT2, THRB, EGLN1, EPAS1*	1p13.3, 15q25.1, 3p24.2, 1q42.2, 2p21	E. Africa, Himalayan Mts, Andes Plateau
Trypanosome resistance	*APOL1*	22q12.3	W. Africa
Malaria resistance	*G6PD, HBB, GYPA, GYPB, GYPC*	Xq28, 11p15.4, 4q31.21, 2q14.3	Tropical Africa, Mediterranean, ME
Toxic arsenic	*AS3MT*	10q24.32	Argentina
Increased BMI	*CREBRF*	5q35.1	Melanesia
Height (shorter)	*CISH, DOCK3, STAT5, HESX1, POU1F1*	3p21.2, 3p21.2, 17q21.2, 3p14.3, 3p11.2	W. and C. Africa
Height (taller)	Polygenic	294,831 SNPs across multiple chromosomes accounted for 45 percent of variance in height.	N. Europe

Note: Adaptation in modern humans is supported by strong genomic and evolutionary evidence. The table shows the type of adaptation and specific genes (where possible) that are associated with it. The chromosomal locations of the genes involved in the specific adaptation are also provided. Note that relatively simple genetic mechanisms are associated with the examples is this table.

Data are modified from S. Fan, M. E. Hansen, Y. Lo, and S. A. Tishkoff, "Going Global by Adapting Local: A Review of Recent Human Adaptation," *Science* 354, no. 6308 (2016): 54–59. SNPs for variation in height are discussed in J. Yang, B. Benyamin, B. P. McEvoy, et al., "Common SNPs Explain a Large Proportion of the Heritability for Human Height," *Nature Genetics* 42, no. 7 (2010): 565–69.

variables such as time and population size. And none of these processes has made for deep separations and variations among people on different continents. This is why we do not consider variation within our species to have created human races.

ARE HUMAN GENETIC DIFFERENCES CLUSTERED INTO GEOGRAPHIC AREAS, OR DO WE CONTINUOUSLY VARY, LIKE A WEATHER REPORT OF TEMPERATURES?

Short answer: humans vary continuously. This is because of a process called *isolation by distance*. The rule of human evolution is continuous variation across geographic areas. Geographic closeness is very highly correlated with genetic similarities.[22]

The first thing to know is the history of human occupation of the globe. We are an amazing species in our ability to occupy virtually all land masses.

Modern humans, those individuals who led to us, arose in and left sub-Saharan Africa sometime between 50,000 and 100,000 years ago. They might have traveled by foot in multiple waves, and some might have left Africa by different routes.[23]

These modern humans encountered other species of humans, such as Neanderthals in southern Europe. Although the groups might have competed and made war, the genetic evidence makes it clear that they also reproduced with each other, and most people of Eurasian descent have a small (1–3) percent of Neanderthal gene sequences (see chapter 9 for more details).[24] The first modern humans arrived in Europe around 43,000 years ago. These people did not contribute much genetic information to modern Europeans. The ancestors of those who currently live in Europe arrived around 7,000–10,000 years ago. Those first modern humans who arrived in Europe were primarily hunter-gatherers. Later, humans who had developed farming replaced the hunter-gatherers. A third wave of migration into Europe occurred about 4,500 years ago (during the early Bronze Age). These migrants were descendants of hunter-gatherers who originated in what today is modern Russia and the Caucasus Mountains.[25]

In addition to being populated by part of the wave of peoples coming through the Middle East, Asia was home to at least two early waves of migration: first by the ancestors of Australians and the Papuans, followed by other ancestors of East Asians.[26]

The archaeological evidence suggests that modern humans were present in Oceania 47,500–55,000 years ago. The extensive genomic study of Papuan and aboriginal Australians suggests that genetic divergence between these groups might have been driven by climate changes and that the aboriginal Australians were living in a relatively high level of isolation until modern times.[27]

The earliest arrival of modern humans in the Americas is accepted to be about 14,000–15,000 years ago. The widespread occupation of the Americas is associated with the Clovis culture (~12,600–13,000 years ago). It is likely that the genomic divergence of Siberians and the American Indian populations began around 23,000 years ago. This suggests that American Indians moved into the Western Hemisphere earlier than indicated by the dating of archaeological remains; however, by what means and routes this occurred is still uncertain and has generated a considerable degree of disagreement.[28]

The migratory paths of humans around the world led to divergences in their gene frequencies, driven primarily by their population history (particularly genetic drift). The answer to the foregoing question focused on the role of natural selection in producing adaptation to new environmental challenges that were faced by our migrating ancestors, but this amounts to the smaller part of the differences in the gene frequencies we observe in modern humans. Also, as people migrated around the globe, the distance between populations reduced the amount of gene flow between them, allowing natural selection to increase the frequency of local adaptations and for genetic drift to differentiate populations from one another by random chance.

The best way to explain the distribution of human gene frequencies is the *isolation-by-distance model*. This means that the genetic similarity of populations is directly a function of how geographically close they are (fig. 2.3). This figure shows that the value of the F_{ST} statistic between any two populations of humans increases linearly with the distance between them. Thus, sub-Saharan Africans and Europeans are more similar to each other than either is to Amerindians. The distance from Lagos, Nigeria, to Berlin, Germany, is 4,555 km. The distance from Lagos to the eastern portion of North America is more than 18,000 kilometers (via the early human migration route). Eastern Europeans and western Asians would be more similar to each other than either is to sub-Saharan Africans, and so on. This change is also continuous, displaying no gaps or clusters in populations. It is also worth noting that the F_{ST} values in figure 2.3 never exceed Wright's critical value for the existence of biological races (0.25). The maximum value of the regression line is ~0.15.

The figure also shows that the genetic diversity of humans declines with distance from Africa. This results from the founder effect (genetic drift) phenomenon. Each newly founded population contains a subset of the genes of the parental population. As this process continued in a series of founding events, each population continued to lose alleles, becoming less diverse. For example, for the human histocompatibility A locus (HLA), there are 31, 23, 20, 21, and 9 alleles for Africa, Europe, East Asia, South Asia, and the Kolla people of northwest Bolivia, respectively. Researchers believe that the extreme lack of HLA diversity in the Kolla resulted from the original population bottleneck that resulted from the migration of humans to the Americas, not some recent event associated with the colonization of Bolivia by Europeans.[29]

There is some debate concerning the processes that explain the isolation-by-distance pattern we see in humans. In the 1990s, the predominant

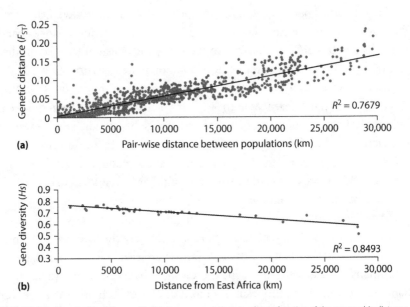

(a)

(b)

FIGURE 2.3. The genetic distances (F_{ST}) between populations is a linear function of the geographic distance separating the pairs (a). The genetic diversity found within a population declines with distance from East Africa (b). *Source*: L. J. Lawson Handley, A. Manica, J. Goudet, and F. Balloux, "Going the Distance: Human Population Genetics in a Clinal World," *Trends in Genetics* 23, no. 9 (2007): 432–39.

explanation for isolation by distance was the serial founder event model described earlier. However, new data resulting from technology that allows for the study of ancient DNA has called into question a simple serial founder events explanation (more details are provided in chapter 9). David Reich and his coworkers suggest that this model will need to be modified to account for remixing of "significantly" diverged populations, or some combination of both models.[30] In either case, a few barriers to migration, such as oceans, mountain ranges, and deserts, modify genetic continuity. But the overriding fact is that human genetic variation is continuous.

WHY DO SOME GENETIC TRAITS CORRELATE WITH EACH OTHER, WHEREAS MOST TRAITS ARE INDEPENDENT (NOT CORRELATED) WITH OTHER TRAITS?

Genetic correlation can result from two processes: linkage or pleiotropy. Of these, the first term is easiest to understand. *Linkage* refers to the fact that genes are found in the same region of a chromosome. For this reason, during crossing over in meiosis (the process that produces egg and sperm cells from the ovaries and testes, respectively), the genes are inherited as a block. If selection favors one gene in a given region, then the nearby genes will also increase in frequency along with the gene under selection. In table 2.1, the genes for adaptation to an Arctic environment (*CPT1A*, *LRP5*) are found at 11q13.3 and 11q13.2, respectively. Furthermore, the genes associated with tolerance to a high-fat diet (*FADS1*, *FADS2*, I) are found nearby at 11q12–11q13.1. Finally, a gene associated with lighter skin pigmentation (which could in turn be associated with better vitamin D synthesis in lower solar intensity) is found at 11q14.3.

It turns out that populations such as the Inuit have high frequencies of the alleles associated with survival and reproduction in the Arctic at all of these loci. This could have resulted from selection that favored any one of them and carrying along all the others. The Inuit managed to live in the Arctic, eat high-fat diets, and evolve lighter skin. The *HBB* gene that confers resistance to malaria is found at 11p15.4. This locus is on the other end of chromosome 11 and is therefore not linked to Arctic tolerance genes. Thus, selection for Arctic tolerance had no impact on malaria resistance.

In *pleiotropy*, one gene is involved in producing multiple traits. For example, having the capacity to tolerate a high-fat diet allows the body to generate greater fat stores, which in turn might produce a greater capacity to retain heat, also allowing for better survivability at Arctic temperatures. This is an example of positive pleiotropy.

On the other hand, negative (or *antagonistic*) *pleiotropy* occurs when a gene influences traits in opposite directions. There is abundant evidence from the literature on experimental evolution that antagonistic pleiotropy played a major role in influencing the survival and reproduction of model organisms (nematodes, fruit flies, yeast). It is likely that antagonistic pleiotropy plays an important role in driving the evolution of genes that cause cancer at a later age (because they are beneficial at earlier ages). It is possible that the increase in frequency of the mutations that causes cystic fibrosis in northern Europeans resulted from antagonistic pleiotropy (beneficial in increasing reproduction and lowering infection, but detrimental in causing cystic fibrosis).[31]

But most of the traits that we see in humans are not at all consistently correlated with one another. This is called *discordance*. Traits that are found on different chromosomes (or arms of chromosomes) will be inherited independently (such as the TYR 11q14.3 and HBB 11p15.4 genes). If these genes influence different traits, there is no reason that they would be genetically correlated. In addition, the various aspects of environment faced by human beings are not always correlated. For example, latitude (solar intensity) and altitude are not correlated. The altitudes of Ethiopia and Tibet are around 4,550 and 5,000 meters (9.14° N and 31.86° N), respectively. Both Ethiopians and Tibetans have high-altitude adaptations but differ in skin color as well as biometric proportions of their skeletons. Thus, Ethiopians are a dominant force in world long-distance running due to their combination of high-altitude adaptation, biometrics, and culture. But no such dominance is observed in Tibetans, because although they have high-altitude adaptation, their biometrics and culture were not conducive to success at long-distance running. However, if you were attempting to climb Everest, you would be better off with a Tibetan Sherpa helping to carry your gear than an Ethiopian.

There is an important implication of discordance; namely, that one cannot extrapolate from one trait to almost any other trait. For example, we might be able to extrapolate from skin color to hair and eye color because of pleiotropy, but we cannot extrapolate almost anything else from skin color.

IS THERE ONE HUMAN SPECIES? CAN WE DEFINE SUBUNITS WITHIN OUR SPECIES? IF SO, WHAT DO WE CALL THEM?

At present, there is only one human species on Earth. You have already learned that this was not always so. Around 300,000 years ago, our species shared the planet with at least six other archaic human species (*H. erectus, H. floresiensis, H. antecessor, H. heidelbergenesis, H. neanderthalensis,* and *H. naledi*). It is unclear whether the Denisovan humans should be considered a subspecies or a species in their own right. The evidence suggests that our species became hybridized through reproduction with the archaic humans *H. naledi* (in sub-Saharan Africa), *H. neanderthalensis* (in Eurasia), and the Denisovans (in East Asia, Melanesia, and Australia).[32] In modern humans, about 1.0 to 2.0 percent of genes originated in these other species.

How one decides to classify human subgroups within anatomically modern humans is not as objective as we might prefer. Modern genomics has allowed the sequencing of thousands of genomes from around the world. The sequence data allow for the application of computational methods to sort individuals into groups. In one popular program called STRUCTURE, one can sort them into any number of groups. But because variation is continuous, some individuals will seem to be made up of more than one group.[33]

However, it is clear that whatever the methods used, no group within our species can be considered a biological race. Interestingly, when genomic methods are used to divide humans into nineteen ancestry groups, it turns out that most, twelve of the nineteen, are sub-Saharan Africans.[34] Why? Because of the greater variation within Africa.

WHY IS IT THAT YOU CAN ALMOST ALWAYS TELL A NIGERIAN FROM A NORWEGIAN, YET A NIGERIAN AND A NORWEGIAN DO NOT GENETICALLY DIFFER THAT MUCH?

Nigerians and Norwegians differ in the frequencies of several physical traits. The most apparent of these are skin and eye color. Yet the genetic foundation of these traits is very simple. There are only five or six major genes associated with skin color and one or two associated with eye color. As there are around 23,000 protein-coding genes in the human genome, only 0.03 percent of the genome accounts for these apparent physical differences.

Also, the differences in allele frequencies along the geographic distance from Nigeria to Norway are continuous. If you were to walk from Lagos to Oslo, you would have a very hard time figuring out where dark skin begins to shade into lighter skin tones. Rather than an abrupt change in skin color or any physical feature, what you'd find is continuous change in the physical appearance of the people you meet.

Yes, if you had a room full of Nigerians and Norwegians you could probably sort them out. But that has nothing to do with "race." Your success in sorting is based simply on the fact that Norwegians and Nigerians live far away from each other and have evolved obvious differences in skin and eye color.

DO OUR SOCIAL CATEGORIES OF RACE MAP ONTO "BIOLOGICAL UNITS"? THAT IS, HOW CLOSE ARE OUR SOCIAL DEFINITIONS AND IDEAS ABOUT RACE TO SOME SORT OF BIOLOGICAL GROUP?

For all the reasons that have been discussed in this chapter, the social definitions of race utterly fail to describe the underlying biological variation found in human beings. In addition, social definitions of race are historically and culturally contingent. The meaning of Black differs in various societies. In the United Kingdom, Black includes Pakistanis and East Indians. In Brazil, it includes many persons of African descent, who would be classified as something else in the United States. In the United States, Black includes anyone with detectable African descent. In the Caribbean, white includes anyone with any European descent, including people who would be classified as Black in the United States. Social races change over time and place. That is fine and appropriate. But science requires stable classifications in order to be replicable. These categories are clearly not biologically justified.

CONCLUSIONS

Human genetic and phenotypic variations are real and measurable. Humans are unusual in that one species has colonized the globe. Yet, humans in different regions have not formed discrete subspecies or biological races. That is a measurable reality. The reasons we have not formed biological races include that we are a young species with little time to develop differences, especially outside sub-Saharan Africa, and the continuous exchange of genes and contact of humans across wide geographic areas.

Human variation is not explained by race. Rather, variation is explained by evolution. Evolution acts locally, not on social races or continents. Race also does not describe human variation. Variation is nonracial. It is almost always continuous, without clear breaks, and it is nonconcordant. Finally, variation within a continent, social race, or local group is much greater than variation among them.

The bottom line is that human biological race is wrong and continued use of the concept in scientific and public discourse is ultimately harmful.[35]

Chapter Three

EVERYTHING YOU WANTED TO KNOW ABOUT RACISM

Is that racist? Am I a racist? What, if anything, is reverse racism?

"Racism" and "racist" are common terms. Not too long ago, racism was unapologetically embraced by the nation's leaders and most of its population. We remember watching Alabama Governor George Wallace on TV as he proclaimed in his inauguration speech on January 14, 1963, to a cheering crowd, "Segregation now, segregation forever."[1] Now, almost no one wants to be called or considered a racist. Donald Trump, despite his public actions, declares himself "the least racist person." Ignoring or being oblivious to racism is obviously damaging. It is also problematic to be paralyzed by the epithet of being called a racist. By neither ignoring nor cowering from the label of racist, we move toward antiracism. And in focusing endlessly on racism as personal, these discussions move us away from some hard work and the focus on racism as a historical and present-day system.

In chapter 1, we outlined the history of the cocreation of race and racism. The preposterous idea of biological races was needed to justify what philosopher Charles Mills[2] calls the "racial contract" and what Isabel Wilkerson refers to as a U.S. caste system.[3] Racism results from the operation of a caste system or levels of closeness to God, the old Great Chain of Being. In this chapter, we move to the present and provide you with basic definitions that help to differentiate among some crucially important but commonly confused terms. We break down the difference between prejudice and racism as well as forms of racism, such as racist actions and thoughts, individual and institutional racism, and how they all connect into a system. We provide clear and concrete examples to help you recognize when ideas and behaviors are racist and when they are not, as well as how individual racist thoughts and behaviors relate to systemic and structural levels of racism.

To be sure, racism is in the ideological air we breathe. We end with a discussion of white supremacy and data showing the persistence of racism and racial inequalities. It is time for all of us to look in the mirror, take on racism, and move together toward antiracism.

WHAT ARE THE DIFFERENCES AMONG THE TERMS "BIGOTRY," "BIAS," "PREJUDICE," "XENOPHOBIA," "ETHNOCENTRISM," AND "RACISM"? HOW ARE THEY DEFINED AND CONNECTED?

These terms are interrelated and have overlapping meanings. *Bigotry* refers to intolerance of any belief or opinion that differs from one's own. For example, one can be a religious bigot. One can be aware of one's own bigotry, or it can be unconscious. *Prejudice*, related to bigotry, refers to an opinion against a group that is typically based on preconceived notions rather than actual experience or reason. It is a preconceived notion that can result from the elementary, logical fallacies of composition and division. The fallacy of composition results when one takes the characteristics of an individual and then infers that all members of a group have that characteristic. The fallacy of division takes a statistical characteristic of a group and infers that all members of the group have that characteristic.[4]

Bias is similar to prejudice and bigotry but slightly more inclusive: it refers to beliefs and actions for or against any object, thing, person, or group compared with another based on preconceived notions. One can also be biased toward an explanation or mode of thinking, and reckoning with such biases is also of great importance. We all have biases and preconceived notions that act as shortcuts. That is, because our biases result from the fact that our brains were produced by natural selection (descent with modification), and many of them are predictable.[5] That is why when we think we've seen something before, our tendency is to take a shortcut to explanation. Guarding against bias is important in science as well as in daily life. One might have a bias toward simple and genetic explanations for racial inequalities in health. These are very common in our culture. Most people prefer simple, genetic explanations for biology and behavior.[6]

Finally, biases might be conscious and explicit. We are aware of our biases. Alan's favorite basketball player is Jaylen Brown of the Boston Celtics. Joe's favorite is LeBron James. Both Alan and Joe love spaghetti and meatballs, one of their favorite suppers as children. But more often, bias is unconscious and implicit. We are not even aware that our brains are pushing us in certain directions because of lessons we learn without thinking of them throughout our lives. When a job applicant is selected over other applicants because he or she has a white-sounding name, that is an implicit bias. The recruiter probably has no idea that he or she is making an unconscious decision based on associations with a name. A great deal of our ideological and personal racism is unconscious or implicit.[7]

Xenophobia is a fear or hatred of others. It appears to be common that some individuals approach those in other groups with some degree of caution. We are, after all, less familiar with the customs and behavior of people in other groups. The Athenians were definitely cultural xenophobes, calling other groups uncivilized or savages, which are xenophobic epithets. Some evolutionary theorists and anthropologists consider fear of strangers (xenophobia) as a sort of hard-wired behavior used for protection of the cultural group. There is evidence for this view, as xenophobia exists in all anthropoid apes (humans, chimpanzees, gorillas, and orangutans) and therefore might have originated in their common ancestor.[8] But make no mistake: even if xenophobia might have some support based on evolution, it is not inevitable, and it certainly is not an explanation for racism. If anything, xenophobia has been used as an excuse for institutional racism. Finally, there is just as much evidence from evolution for the desire of others and the evolutionary importance of exogamy, mating outside one's group.

Xenophobia also lapses into *ethnocentrism*, a preference for one's own culture and the evaluation of other cultures according to preconceptions derived from one's own. However, humans also seem to have an attraction to others. The other is often seen as exotic and attractive. Although there might be some evolved and genetic basis for xenophobia and ethnocentrism, the specifics are highly cultural. For example, Donald Trump's stated preference for immigrants from Norway or Eastern Europe rather than the Middle East or Mexico is his personal, learned expression of his ethnocentrism and xenophobia.

These ideologies and behaviors intersect with one another, as they all have to do with negative preconceptions and behaviors toward individuals in other groups, whether different genders, social classes, castes, ethnicities, citizens of other states or countries, or race. These negative perceptions are often unconscious.

Defining Racism

As we discussed in chapter 1, the most common and simple definition of racism is prejudice plus power. This definition highlights that racism is a form of prejudice against people of another socially defined race. However, to be racist requires a power differential or an ability to move the wheels of institutions that have power over individual lives. By this definition, within the contemporary U.S. power structure, a Black or brown person can hold prejudices against a white person, but they have no or limited power to act on their prejudice and to be racist against a white person. By this definition, reverse racism is not a thing (see the later discussion). The cultural legacy of the United States is responsible for the moronic concept of reverse racism. For most of our history, it was normal for whites to unashamedly enact racist policies and laws against Blacks, browns, and reds.

We like the short definition of racism (prejudice plus power) but want to go further and make sure it is clear how racism is linked to the belief in biological races. The "rocket fuel" of racism is the belief that biological races are different, innate, and hierarchically arranged by God and/or evolution. Ibram X. Kendi says that if you believe in biological race, you are a racist.[9] We would say you are an *ideological* racist or racialist. However, there is a strong correlation between racialism and racism. The belief in innate biological differences can then justify racial differences in wealth and health as just part of those innate differences rather than being attributable to structural and institutional forms of racism. The belief in biological race can become an excuse for racism.

To clarify, there are two intersecting forms of racism. One is more personal and ideological and includes the varied and everyday notions of racial difference and hierarchy. It is what individuals think, consciously or subconsciously, which form patterns of thought and actions that are inherited through history and culture. Kendi says that this is a simple descriptor of "racist." Most people harbor racist ideas. In the words of Beverly Tatum, we all breathe the ideological smog of racism.[10] However, a primary purpose of this book is to maintain that it is possible to cure oneself of racist ideology. It does take work, however, as so much in our society continues to reify racial thinking and to reward racist behavior.

Personal or individual racism is important because without it, the second and most important form of racism—institutional and structural racism—would lose ideological support. In her bestselling book *White Fragility*, Robin DiAngelo makes the important point that racism is not about individuals, it is not an action, and it does not require intent. Rather, racism is systemic and institutional.[11] We agree that the racism that affects human lives is systemic and institutionalized. And we think DiAngelo would agree with our addition that individual-level racism—especially the worldview of innate, unchanging, and hierarchically arranged biological races—however implicit and buried it might be, is the ideological fuel for institutional racism. That kind of racism in employment, for example, rests on the many managers who, time after time, reject applicants based on Black-sounding names. You might not think you see color, but we all do. Racism is like oxygen in the air we breathe. You might not see it, but science can measure it.

ARE THERE DIFFERENT FORMS OF RACISM?

Yes, racism takes a variety of forms with different central elements. One typology developed by sociologists identifies four prominent, intersecting types: biological racism, symbolic racism, ethnocentrism, and aversive (color-blind) racism. In addition, as noted earlier, racism can be individualized (interpersonal and internalized) and institutional. Racism can also vary in being entirely intentional to entirely unintentional. All of these types of racism intersect and can support one another, so they are all part of a system. They all contribute to suffering and limit human and societal potential.

Biological racism rests on the premise that races exist in the human species and that these races differ in their innate (genetic) capacities. It purports that the social status of these groups results not from discrimination but from their innate capacities. Among these innate capacities are traits such as intelligence, morality, and longevity. Thus, biological racism posited that certain races are more likely to produce criminals than others, providing a handy justification for the preponderance of African American convict laborers in the 1930s and the mass incarceration of African Americans to the present day. In the late nineteenth century, this ideology also predicted that "Negroes" would go extinct due to their lack of "persistence." Frederick Hoffman predicted, in *Race Traits and Tendencies of the American Negro*,[12] that the Negro would become extinct because he was unable to adapt to the rigors of northern cities and American civilization. We believe that biological racism is the foundational racism, because all the other forms of racism rely on the notion the underlying biological differences exist between racialized groups of human beings.

Symbolic racism is a form of prejudice held by individuals of European descent against those of African descent. Other American ethnic groups also adhere to this belief system, being prejudiced against any group that is different from their own. Symbolic racism is usually described as a coherent system that can be expressed in several beliefs: that individuals of African descent no longer face serious prejudice or discrimination, that their failure to progress results from their unwillingness to work hard enough, that they make excessive demands, and that they have received more help from the government than they deserve.[13] Symbolic racism feeds into biological racism.

Ethnocentrism is the tendency to evaluate other ethnic groups by the standards of behavior and qualities displayed in one's own ethnic group. Judging standards of beauty by physical traits inherent to one's own ethnic group (hair type and skin color, for instance) is one example. Judging another group's patterns of expression as vulgar, based on how one's own group expresses itself, is another. Ethnocentrism in and of itself is not a form of racism, because it does not always involve a power dynamic. However, ethnocentrism is aligned with symbolic racism, and with the added element of power and belief that differences are innate, it leads to racism.

Cultural racism is the belief that different races have different cultures that lead to particular outcomes, such as better education, more wealth, and a better society. It is an effort to separate from biologizing differences. There are three fundamental problems with the notion of cultural racism. First, it is a sort of oxymoron. As we discuss, a fundamental aspect of race and racism is biologization. Linnaeus was culturally racist, in that he confused cultural and biological traits in his racial classifications, thinking they were all essential to racial types. But today, few have this problem. Second, just as there is no essential whiteness, there is no essential white culture. Cultures, by their very definition, are constantly mixing and changing. Finally, cultures do not form in vacuums. For example, one cannot know or evaluate a health behavior, diet, exercise, and the like without thinking about housing and access to healthy foods. Those things are cultural but also consequences of regional, national, and global flows of people, ideas, and things.

Color-blind or aversive racism is an ideology that allows those of the dominant socially defined race (those of European descent) to claim that racism is no longer the central factor determining the life chances of people of non-European descent (particularly dark-skinned individuals of African descent). This position argues that instead of the ongoing institutional and individual racism of American society, nonracial factors such as market dynamics, naturally occurring phenomena, and the cultural attitudes of racial/ethnic minorities themselves are the main causal factors of their social subordination. Indeed, recent studies have shown that although there is near universal endorsement of racial equality as a core value, aversive racism persists.

People practicing aversive racist behavior would never describe themselves as racists, but this form of subtle, indirect racism operates across a wide variety of settings, such as in employment, legal decisions, group problem-solving, and everyday helping decisions. An example of an everyday helping decision is whether or not a white person stops to hold a door open for a person who is not white, or which persons someone decides to help, such as when white rather than nonwhite homeless people are given more donations of money or food.

Our culture tends to focus on individual racism, that is, the racism that individuals undertake as actions, behaviors, and underlying racist thoughts and ideas that determine their behavior. Recent studies have found that all humans harbor unconscious stereotypes or implicit biases. When most of us think about racism, we think about individual racism. And, yes, individual racism is important.

The main importance of individual racism is not just in how it impacts one's thoughts and actions but in how it ramps up into institutional racism. When individual ideas become socially agreed "fact," powerful institutions can act to discriminate. Take redlining as an example. It became a widespread practice in the United States to deny loans and housing to families of color, and this practice of redlining led to different capacities for accumulating wealth. Specifically, redlining influenced who was able to receive loans to buy a home. Considering that homeownership has been seen as one of the avenues by which wealth is generated and transmitted to the next generation, redlining explains, more than any other recent racist practice, the difference in wealth between whites and Blacks in the United States.

In 2014, the median household income for whites was $71,300, compared with $43,000 for Blacks. The difference is almost the same when we control for education. For college-educated whites, median income was $106,600 compared with $82,300 for college-educated Blacks. However, worse than just the household income difference is the disparity in overall wealth, which includes all assets a family owns, including stocks, bonds, and properties minus outstanding loans and other debts. The median wealth in 2016 was $13,204 for Blacks and $149,703 for whites—a ratio of 1 to 11.5—and has not changed since 1968.[14]

Finally, it is easy to recognize the violent racism of slave owners and Nazis. If they were the only racists, racism would be a problem that we could more easily isolate. Unfortunately, the ideology of race as biological and hierarchical has permeated society for so long that we hardly notice how pervasive it is. As we've said, it is reified. The doctor who refused to diagnose scleroderma and the physician who failed to give orders for a bone density test (see the introduction) are not overt racists. Rather, they are following their medical training, which is racialized in ways that disempower individuals and communities of color.

Consider the idea of mean and kind racists. Mean racists are those who recognize their hate and intentionally do harm.[15] Kind racists include pretty much everyone else, who might see racism as an evil yet fail to fully recognize the humanity of individuals of different races. Kind racists support mean racists, and both contribute to systemic racism.

WHEN DID RACISM BEGIN?

Racism and race began together. As race evolved from a folk belief to a legal entity to a set of pseudoscientific facts, both institutional and ideological racism became more established. Racism requires race, and biological race provides intellectual cover for racism.

Given that formulation, racism cannot have existed before race. The treatment of Jews throughout medieval Europe and their expulsion from Spain in 1492 are acts that presage racism.[16] These acts were ethnocentric, intolerant, and bigoted. The belief that a Jew will always be a Jew, that Jews cannot change their essential Jewishness, and that Jews differed physically from Christians is close to believing Jews are a distinct race. But these actions certainly are not full-blown racism as at this time Jews were not fully considered to be a race.

The enslavement of Africans and the start of the triangle trade was an indication of the beginning of full-blown racism. Enslaved Africans were thought to be subhuman and of an entirely different type of human. They were commodities that could be enslaved, traded, owned, and put to death. Slavery is an economic system that is supported by the ideology that the enslaved individual will always be of less value than the slave owner. The enslaved person was either degraded (monogenism; e.g., the mark of Cain or curse of Ham) or created separately (polygenism; e.g., pre-Adamite races).[17]

Chattel slavery is often pointed to as the essential form of institutional racism. We have no argument with that. However, in the seventeenth and eighteenth centuries, ideological racism had not completely developed. Race was not fully reified. Theologians, scientists, and politicians struggled to justify their belief that Native Americans and Africans were less than Europeans. One sees evidence of this ideological struggle in the words of Darwin, Jefferson, and many other influencers of their time.[18]

Jefferson, in *Notes on the State of Virginia*, pondered the ethics of slavery.[19] He found himself in the contradictory position of writing in support of liberty yet owning slaves, not to mention his sexual relationship with Sally Hemings. As we noted in chapter 1, in *The Voyage of the Beagle*, Darwin made a number of comments on the gap between the English gentlemen and native South Americans, but he also said, "If the misery of the poor be caused, not by the laws of nature, but by our institutions, [then] great is our sin."[20] Jefferson held to the idea of the inferiority of Africans but was more sympathetic to the notion of Amerindian equality with Europeans.[21] On the other hand, the young Darwin was raised in an abolitionist family. His grandfather, Josiah Wedgewood, designed the British Anti-Slavery Society medal.[22] Darwin, unlike Jefferson, later in his career would contribute major scholarship that helped to debunk the polygenist views of this period concerning the existence of separately created species and an innate hierarchy among human types.[23]

Perhaps it is surprising that we observe that racist ideology—the ideology of different races hierarchically arranged—might be as great now as at any time in history. That is because the *idea of race* is at full strength today. Since at least the time of Jefferson, a counter ideology of egalitarianism and democracy has tried to chip away at the ideology of racism. Racial equality is in the Constitution. A plethora of laws prohibit discrimination by race. Scientists have even proven that our species does not have biological races. Evolution is the main scientific discipline that demonstrates that biological races do not exist within in our species. Yet, in the United States, the majority of Americans either do not accept that evolution is true or simply do not understand it.[24]

A significant sector of American society draws its belief concerning human biological variation from religious teachings, particularly special creationism. The racial attitudes of religious denominations in the United States vary considerably, with some of the most racist beliefs and behaviors exhibited by evangelical Christians.[25] For example, in 2020, Quan McLauren, the diversity and retention director at Liberty University, resigned, citing university president Jerry Falwell Jr.'s racist and oppressive leadership.[26]

Throughout our careers, we have been fighting an uphill battle against racism. That is because the ideology of race and racism has history and power on its side. In table 3.1 we visualize the history of racism in America as occurring during a single day. Laws that attempted to chip away at the lived experience of racism didn't start appearing until late in the afternoon. The most significant of these laws did not appear until after 8:00 P.M. Eroding biological racism is difficult, because when scientists present their findings concerning biological variation, most individuals think they see race when what they actually see is skin color variation. Race is reified, and institutionalized white supremacy rewards racist behavior. This is exactly the ideological fuel needed to keep America's racial hierarchy in place.

TABLE 3.1
African American Social Experience Presented as a Single Day

Event	Year	Clock Time
First Africans in Jamestown, VA	1619	12:00 A.M.
Virginia laws differentiating Africans from other servants by race	1682	4:47 A.M.
Fugitive Slave Act	1850	1:50 P.M.
13th Amendment ends slavery	1865	2:43 P.M.
Plessy v. Ferguson—separate but equal	1896	3:33 P.M.
Red Summer—race riots massacre of African Americans, Tulsa massacre	1919	5:57 P.M.
Black vs white incarceration ratio is 4:1	1950	7:49 P.M.
Emmett Till lynched	1955	8:07 P.M.
Mass incarceration ratio reaches 5:1	1960	8:25 P.M.
Civil Rights Act	1964	8:29 P.M.
Voting Rights Act	1965	8:42 P.M.
Mass incarceration ratio reaches 6:1	1970	9:01 P.M.
Joe Graves earns PhD in evolutionary biology	1988	10:03 P.M.
Mass incarceration ratio reaches 7:1	1989	10:04 P.M.
Killing of George Floyd, Breonna Taylor, Ahmaud Aubrey	2020	11:59 P.M.

WHAT ARE EXPLICIT AND IMPLICIT BIASES?

Quotes such as "I'm not racist," "I treat everyone the same," and "I don't see color" are commonplace. Making fun of these color-blind notions, late night comedian Steven Colbert said, "I don't see race. People tell me I'm white, and I believe them."

Explicit biases refer to the attitudes and beliefs that people have about a person or group on a conscious level. Much of the time, these biases and their expression arise as the direct result of a perceived threat. For example, a white woman might say that she wants a white female teenager to babysit her child because she feels more comfortable with a white girl than a Black boy. But more often, these biases are unconscious. She might not consciously realize why she picked the white girl, or that she crossed the street to avoid coming close to a Black male.

As a graduate student in 1981, Joe had a door slammed in his face as he attempted to enter the University of Michigan's Museum of Zoology. The woman who slammed the door assumed that as a Black male, he had no legitimate reason to be in the building on the weekend. When he produced his key and entered the building, she continued to interrogate him about his purpose for being there. She assumed that there were no Black graduate students in evolutionary biology. He explained to her as gently as he could that he was going back to his laboratory to take care of his animals.

Implicit biases are unconscious attitudes, beliefs, and stereotypes that affect our understanding, actions, and decisions. We all have unconscious biases. They help us to make decisions. From birth, our brains collect information. We come to have expectations about what is safe to eat, what is a soothing sound, and what a friendly face looks like. We also tend to classify individuals. We grow up to associate lots of behaviors and characteristics with females and males. And we apparently do the same based on skin color and race.

Project Implicit, the developer of tests of implicit bias, starts with the example of someone who smokes a pack of cigarettes a day.[27] One could hide this information because of embarrassment. That is an explicit bias; it is a purposeful deceit. On the other hand, one might not be aware of the amount that one smokes and thus might underestimate it. That is implicit bias.

Studies of implicit racial biases have recently mushroomed, uncovering the breadth and depth of such bias and how it influences decision-making and actions in education, health care, employment, and law enforcement. Implicit racial bias, we now objectively know, is everywhere.

One early example of implicit bias concerned physician recommendations to patients who reported heart problems.[28] The researchers recruited white and Black cardiologists at a medical conference and had them observe four actors who were trained to act like patients: a Black female and male and a white female and male. The researchers asked the cardiologists to recommend tests and treatments. The authors found that both Black and white cardiologists treated the Black patients less aggressively than the white patients. After being asked about their different diagnoses and treatment suggestions, the physicians discovered that they were unaware that they had made them.

Similarly, a 2016 study examined interactions between Black patients and white oncologists who had been administered a test for their levels of implicit bias.[29] The authors found that oncologists who rated higher in implicit racial bias had shorter interactions with their patients, and their patients rated the interactions as "less patient-centered and supportive" than doctors with less implicit bias. The study also found links between a physician's bias level and their patients' confidence in the physician's recommended treatments, as well as more perceived difficulty in completing them. These and other examples of implicit bias might contribute to decreased effectiveness of health care of Black patients and the huge racial inequalities in disease and death rates.

One of the most powerful examples of the harm caused by implicit bias comes from its role in determining death sentences.[30] Previous research on homicide sentencing had shown that when the offender was Black and the victim was white, the offender was more likely to receive the death penalty. Stanford professor Jennifer Eberhardt and her team examined more than six hundred death penalty cases from Philadelphia, Pennsylvania, from 1979 to 1999. They found within this dataset that the more the offender displayed stereotypically "Black" physical features, the more likely they were to receive the death penalty.

Consider the difference between David Duke, the former grand wizard of the Ku Klux Klan, and Donald Trump. Duke is an explicit racist. He does not hide his hopes for unequal and as separate as possible. Everyone knows where he stands. He is an explicit white nationalist and white supremacist. Conversely, Trump claims to be "the least racist person." Let's assume that he believes that to be true. If so, then his racism is implicit. His words and actions, including his association with and failure to speak out against white supremacists like Duke, his enacting of the Muslim ban, and his support for building a border wall, make clear that he implicitly fears Black and brown people.[31]

To summarize, biases, whether explicit or implicit, are important because biases or preconceived notions directly influence actions. These biases have been shown to influence who gets hired for a job.[32] The résumés of individuals with traditional Black names are rejected more often than those with traditionally white-sounding names. These biases constrict opportunities from the cradle to the grave. They contribute to institutionalized racism because these biases are deeply cultural and get stuck in the minds of those with the greatest political power.

AM I A RACIST IF I... ?

Questions that start with "Am I racist if" are, in truth, a little off. As we've said, racism is not so much about individuals as it is about a system. Racism is not so much about thoughts and personal behaviors as it is about histories and institutions. That said, the question "Am I a racist?" is a common one and important to answer. If you are reading this book, we assume you do not want to be a racist.

Nobody is purely a racist or an antiracist. We all, one hopes, are striving to be more antiracist. But sometimes we have racist thoughts and, worse, act in ways that perpetuate racism. Our parents probably repeated racial stereotypes, which got stuck in our brains. We live in racial smog. And our thoughts and actions are connected to systemic, institutional racism. As a house is built on brick and beam, so a system of institutional racism is built on racist idea and racist action.

Yes, we all harbor some racist thoughts. Some might be explicit, and most are probably implicit. We live in a society and time that make it impossible to not be infected by racist memes and not breathe the air of racial differences, codes, and racism. All of us—but perhaps especially folks with so-called white skin privilege, like Alan—need to be aware of them, to be sensitized to them, and then to call them out. Do not be afraid to address them.

Actions that typically come from unconscious thoughts can do harm. Denying a job interview because of a Black-sounding name might be an unconscious thought that leads to an action. It is not yelling and threatening to call the cops on a Black bird-watcher, and it is not a policeman with a knee on the neck of a Black man. Those actions are different. Actions arising from unconscious thoughts probably will not make the evening news. However, such everyday racism as denying a job is one of the myriad hidden bricks of systemic racism.

If you do something that, in retrospect, is racist, we recommend that you consider the difference between an act of racism and being a racist. A single act does not make you a racist. As Ibram Kendi says, we all do racist stuff.[33] For our culture, that is a norm. The point is to be more aware and, we hope, move from more to less.

Now, why is this all a little off? If you've been reading other sections, you might know what we are about to say. Pause. It is because being racist is not about a specific action. Rather, it is about a system and operating within a system that perpetuates racial inequality.

IS THERE SUCH A THING AS KIND RACISM?

The distinction that historian Donal Muir[34] makes between kind and mean racism is close to the distinction between conscious (explicit) and unconscious (implicit) racism. Muir argues that few individuals are intentionally racist. The policies of the KKK are mean, explicitly racist. Nazi laws and genocide were explicitly racist. Jim Crow laws were explicitly racist. The banning of Colin Kaepernick from the NFL was most likely explicitly racist because it was planned—perhaps not formally but nonetheless planned—as a punishment for his activism.

But as Muir argues, most racism is not mean. Perhaps it is not exactly kind either, but that is his term. This racism is often paternalistic. It is thought to be helpful, or at least not harmful. For example, dividing research subjects into races might be seen as kind because it provides separate results by race. If the researcher is focused on genetic differences, it is also racist in thinking that races might respond differently to a drug or treatment. It would not be racist to recognize that socially defined races experience different social, cultural, and physical environments and therefore might respond differently to a drug or treatment. In the former case, the researcher is assuming a biological, innate, and immutable difference between these groups. In the latter case, the researcher is concerned with how institutional racism harms racialized individuals.

Kind racism is also, well, sort of kind. It is the complement of "oh, Blacks have such great rhythm" or "the Asians are so hard working" or "those Jews are great with money." Kind racisms leave the door open for mean racism.

IS USING THE "N WORD" RACIST? IS IT RACIST IF I AM WHITE? IS IT RACIST IF I AM BLACK?

Yes, the N word is racist if you are not Black. No, the N word is not racist if you are Black. However, the continued use of this word by African Americans has links to deep self-hatred that was internalized through years of racial subordination in America.[35]

The N word is a virulent epithet. It harks back to the time of slavery and the Jim Crow era. It harks back to the Negroid race and racial rankings. It is pure hierarchy. It has hurtful overtones. It is not unlike epithets for other minorities and ethnic groups. We need not repeat them here. The N word is at the far end of a continuum of hurtful epithets. Let it die.

If you are Black, uttering the N word might not be polite. It might still shock, and it might still hurt. But it is also a way to defang the short word by embracing it and making it one's own. Fair enough. It is for Blacks only. But Joe would say that Black people will never be truly liberated until they forever erase this word from their vocabulary.

IS ANTI-SEMITISM A FORM OF RACISM?

Yes, and it is a unique form of racism. Anti-Semitism is its own unique form of intolerance and hate and, at the same time, a form of racism.

First, a bit on the illuminating history of the term *anti-Semitism*. From the German *Antisemitismus*, the word was coined in 1879 by William Marr, a German political agitator, to replace *Judenhass* (literally "Jew hatred").[36] The linguistic move was to hide hatred of Jews behind the façade of science. With anti-Semitism, Jews became a Semitic people, or the Semitic people, a race or subrace, and hatred of Jews could then be fit into a scientific hierarchy, Nordics and Aryans at the top and Semitic peoples near the bottom. Anti-Semitism was rationalized.

Some have debated whether anti-Semitism remains the appropriate term for hatred of Jews. Its advantage, racializing Jew hatred, is also its disadvantage. It continues to highlight the old trope that a Jew is always a Jew. In the following, we use "anti-Semitism" or "Jew hatred" when one or the other seems most appropriate.

Why, then, is Jew hatred different from racism? The answer is that Jew hatred does not fit the classic definition of racism: prejudice plus power.

Jews experience lots of anti-Jewish prejudices and hate. Jews have been victims of hate longer than any other group. But unlike other oppressed groups, Jews are seen as both less than (dirty, disgusting, infectious) and as a powerful cabal.

A common trope of Jew hatred is that they are inordinately powerful, somehow in control of the media, banks, and other industries. This lie of Jewish power dates from the *Protocols of the Elders of Zion*,[37] a fabrication produced in Russia before the revolution, widely circulated by Henry Ford in the 1920s, and still frequently reproduced. This myth of Jewish power is articulated on the political left, such as in the *Sociology of Freedom*, the 2020 book by Abdullah Öcalan,[38] the imprisoned leader of the Kurdistan Workers' Party. Öcalan writes, without citations, about an eternal and essential Jew with a powerful ideology. Given Öcalan's stature on the left, his words gained credibility. A widely circulated lie takes on a life, it continues to act. And this leftist anti-Semitism feeds the right and rightists' chant that the (powerful) "Jews will not replace us" (with Black and brown people).

However, Jews are neither powerful, as they have been thought to be in imaginary works, nor do they have an ideology. Jews differ from place to place and time to time. Under Christianity, Jews gravitated toward jobs and positions that were open to them. But there is no Jewish cabal. Believing in such is Jew hatred and lacking in facts. Jews are thought to be both communists and, at the same time, virulent capitalists, and hated for both. Jews are what whites imagine them to be.

Here are some facts about Jews. The number of Jews alive today is less than fifteen million. (It was more than sixteen million before the Holocaust.) More than half a million Jews reside in just two countries: the United States and Israel. Jews make up about 1.8 percent of the U.S. population. That's because Jews escaped from Jew hatred in Europe and the Middle East to Israel and United States. That's hardly the stuff of worldwide control.

Judaism is a religion practiced by people of varied genetic ancestries. That's confusing to many who blanketly call Jews white or think that all Jews have white skin privilege. Many Jews do indeed pass as white and have white skin privilege. These Jews, including Alan, are eastern European or Ashkenazi. Development of the Jewish diaspora led other groups to join, including Sephardic Jews of Iberia, northern Africa, and locals throughout Asia, the Mediterranean, and Africa. The stereotypical Jew is Ashkenazi. After World War II, they became categorized as white, perhaps honorary whites or whites of a different shade, as Jacobson aptly describes[39]—but, most important, white.

And although not all Jews are middle or upper class, many are. They have some economic power. The result is that on measures of impact of racism, such as education and economic status, Jews in the United States and most European countries seem to be doing okay. Once they could not gain acceptance into certain country clubs, live in certain locations, be admitted to certain hotels, or go to certain schools (like Harvard), but now they can, explicitly if not always implicitly.

Why, then, is the answer to the question, "Is anti-Semitism racism?" a yes? As has been clear since the Middle Ages, Jews have been persecuted. Throughout history, they have been victims of religious persecution, from their expulsion from Spain, to pogroms throughout Eastern Europe, Hitler's Holocaust, and the rise of anti-Semitic tweets, desecrations, and murders in the United States. These add up to ethnic and racial discrimination and more.

The "more" is captured in stereotypes. The Jew is often seen as a racial type, as having inherent qualities that are biological. In Spain, a Jew would always be a Jew. In Nazi Germany, Jewish blood needed to be avoided at all cost and Aryan blood needed to be protected. Science got involved to study the Jewish type. Efforts were made to recognize Jews and to explain how they became the way they were. And these efforts supported their eradication. As they often do to support racist institutions, science and law worked together.

In a research report published in the *American Journal of Physical Anthropology* in 1929, Manoiloff, a Russian scientist, wrote that he could discern Jewish from Russian blood. He declared that Jewish blood was paler than Russian blood, which also contained more adrenaline. Manoiloff added reagent to the blood of known Jews and Russians and was able to deduce that they changed color so he could with almost perfect accuracy detect Jewish from Russian blood.[40] Of course, this is the quintessential unrepeatable experiment. We do not know what reagent he used, and the results are preposterous. But, as Hitler was soon to seize power, it would be important to discern who is a Jew, and the thought is that it was in the blood.

Anti-Semitic science flourished on both sides of the Atlantic. At the University of Virginia Medical School, Robert Bennett Bean, head of the Department of Anatomy, joined a chain of scientists who were fascinated by Jewish noses. His 1913 paper in the *Anatomical Record*, a well-respected publication, is devoted to the study of the unique characteristics of the Jewish nose, complete with an analysis of the surrounding muscles and a theory about which factors accounted for how the Jewish nose form evolved.[41]

Given how much anti-Jewish racism exists today, we recommend letting the term "anti-Semitism" die and instead use either "Jew hate" or "anti-Jewish racism." Because that's what it is.

IS ISLAMOPHOBIA RACISM?

Similar to the answer regarding Jew hate, Islamophobia—fear or dislike of and prejudice against individuals who identify as followers of Islam, or Muslims—is and is not a form of racism. Islamophobia has not yet taken on the same degree of racist tinge as anti-Semitism. Fear of such individuals focuses more on Islam as a culture, religion, and political power than as a racial essence. However, Islamophobia is a form of virulent ethnocentrism that could be considered a form of cultural racism, that being a Muslim is culturally essential.

In 2019, the Institute for Social Policy and Understanding reported that Islamophobia in the United States had increased from the previous year. The increase differed by socially defined race, ethnicity, and religion. Jewish and Hispanic Americans had the most favorable views of Muslims, whereas white evangelicals had the least favorable view (44 percent favorable versus 20 percent unfavorable). On the other hand, Jews had greater than five times more favorable than unfavorable views (53 versus 13 percent), Hispanics had about five times more favorable than unfavorable (51 versus 10 percent), and African Americans seven times more favorable than unfavorable views (35 versus 5 percent).[42]

At the level of politics, our fear is that the attempt of former President Trump to guard the borders and try to ban travel for individuals from so-called Muslim countries feeds into existing misrepresentations and fears of Muslims. It seems to us that Islamophobia at the political level generalizes from cultural and political intolerance to stereotypes about more than one billion followers of Islam. That is certainly prejudice, and with differential in power, it is also a form of racism.

PEOPLE KEEP TALKING ABOUT "REVERSE DISCRIMINATION," BUT WHAT IS IT?

Nothing.

This one is easy. Reverse racism, or reverse discrimination, is not a thing. It is a myth. The idea of reverse racism refers to the assumed overreach of affirmative action programs that are aimed at equalizing past injustices against minorities of color. But that is not true.

Now for some history. The idea of affirmative action effectively began in 1961, when President John F. Kennedy issued an executive order creating the Committee on Equal Employment Opportunity. This very limited order called only for all hiring programs supported with federal funds to ensure that they are free of racial bias. A year after the passing of the Civil Rights Act of 1964, President Lyndon Johnson framed affirmative action this way in his 1965 Howard University commencement address:

> You do not wipe away the scars of centuries by saying: "now, you are free to go where you want, do as you desire, and choose the leaders you please." You do not take a man who for years has been hobbled by chains, liberate him, bring him to the starting line of a race, saying, "you are free to compete with all the others," and still justly believe you have been completely fair. . . . This is the next and more profound stage of the battle for civil rights. We seek not just freedom but opportunity—not just legal equity but human ability—not just equality as a right and a theory, but equality as a fact and as a result.

From the 1960s to today, the idea of affirmative action has been hard to enforce, widely debated, and widely misunderstood. The most important legal case to challenge affirmative action is known as *Regents of the University of California v. Bakke*. In this 1978 case, Allan Bakke claimed that he was denied admission to medical school because of his race. He is white and challenged that positions in the entering class of Davis Medical School had been set aside for disadvantaged minorities, thus disadvantaging him.

But Bakke had the advantages of his white skin privilege. This did not go away. All that was being tried was to make the advantage a bit less.

What Johnson was aiming for over a half century ago was not equality, that everyone is given an equal chance but starts from widely different places. Rather, he was trying to take small steps toward equity, that everyone has an equal chance of admission to school and a job and an equal ability to own a home and build equity in it.

CAN ANYONE BE A RACIST? CAN A PERSON WITH LITTLE ACCESS TO POWER BE RACIST?

No. Like the answer to the reverse racism question, this one is easy. Again, we are helped by the useful definition of racism as prejudice plus power. By definition, those without power cannot be racist. They can—and are often—prejudiced. We all hold some prejudices. Prejudices are cognitive shortcuts. Without power, you can think like a racist, you can be a cognitive racist, but you cannot set in motion the levers of institutional racism. Your racism does not act on the world.

The most important thing to take away is that racism is not about individual skin colors but about all of our silent collusions and complicities in institutions that change lives.

IS WHITE SUPREMACY ON THE RISE, AND IF SO, WHAT'S GOING ON?

First, let's look at the question, "Is white supremacy on the rise?" Our answer unfortunately seems to be yes: white supremacy is on the rise. It is doubtlessly increasingly visible, and more and more individuals are joining white supremacist groups. The harder question is, "What's going on?" There are a number of theories and ideas about why white supremacy is on the rise and what is in the heads of people who think white supremacist thoughts and act in supremacist ways.

The U.S. State Department has tracked the rise in hate crimes. The Anti-Defamation League tracks anti-Semitic incidents and found that they reached an all-time high in 2019, the last year for which the organization has data. Similar increases have been seen in hate crimes against Black and brown people. The most egregious of these were the 2020 killings of Ahmaud Arbery, Breonna Taylor, and George Floyd. Many of those who stormed the Capitol building on January 6, 2021, held white supremacist beliefs and belonged to white supremacist militias.

Our sense is that whites—especially working-class whites—feel increasingly threatened by Black and brown people. They feel that they have less control over their lives. As we write, COVID-19 is still rampant in the United States (and elsewhere in the world). There are huge political divides in America. We live in uncertain and stressful times. It is easier to blame the loss of control on scapegoats, so Black and brown people are portrayed as taking jobs from whites and somehow robbing others of the American dream. This situation has unsettling parallels with how anti-Semitism was used as a catalyst to fuel the growth of the Nazi movement in Germany. On August 11–12, 2017, there were more white supremacists marching in Charlottesville, Virginia, than there were fascists marching with Hitler in the Munich Beer Hall Putsch of November 1923. We believe that in the current situation, we must pay careful attention to the growth of white supremacy and the hate that it generates or we risk being caught off guard, as were Germans in 1933. We also warn that the lack of a strenuous response to the storming of the Capitol has actually emboldened the white supremacist movement.

WHAT IS WHITE FRAGILITY?

White fragility is a framework popularized and used by Robin DiAngelo to describe a common pattern of behavior and feeling expressed by whites when engaging with race and racism.[43] DiAngelo writes that white people are protected from dealing with race and racism. They are insulated by their majority status and the sense that white is not a race, the result of growing up with the expectation that race is not their problem and thus lowers their ability to endure racial discussions and deal with racial situations. As a result, they are frail.

White fragility is a condition in which any degree of racial stress becomes overwhelming and intolerable and thereby triggers a defensive posture. Instead of dealing with hard discussions about the racial world order, whites often say "I am not racist" or "That's not me" and change the subject.

DiAngelo believes that white fragility is a major impediment to becoming antiracist.

WHAT ARE THE DIFFERENCES AND SIMILARITIES BETWEEN RACE AND CASTE?

Most scholars view a racial contract and castes as different and parallel forms of social and politically sanctioned hierarchies. A few scholars and writers, most notably Isabel Wilkerson and Michelle Alexander, argue that the Western racial hierarchy is a form of caste.

One's caste in India is determined by birth and is permanent. Intermarriage among castes is discouraged, if not prohibited. This has resulted in genetic divergence to occur between the castes to the extent that they can be identified through genetic markers.[44] Caste is like class but far more immutable. In India, Dalits are the lowest caste, previously known as "untouchables."

In the United States, race is associated with class hierarchy and is certainly caste-like. This is an old observation. Both Swedish sociologist Gunnar Myrdal and British anthropologist Ashley Montagu made this observation in the 1940s.[45] Caste is like race, in that it is assigned at birth based on the caste into which one is born. It is like a class system, in that social immobility leads to a trap that makes it hard to escape one's class and caste. Liberal thinkers might say that class mobility is possible, whereas caste mobility, moving out of one's caste, is not possible. However, the data on low class mobility in America refutes the notion that these are so different. Sociologists have long recognized that capitalist economies operate with dual labor markets.[46] In America, racially subordinated groups have been differentially relegated to the secondary market, which is characterized by menial and irregular labor.[47] To make the situation even worse, in the later portion of the twentieth century, the irregular labor market grew, as well as the percentage of people who are structurally unemployed.[48] This means that upward social mobility for most racially subordinated persons in America is gradually ceasing to exist.[49]

Wilkerson recently wrote that race in America is much like the caste system of India. One is born into one's race, and the expectations and limits of race are the same as in a caste. She formulates that race is a visible manifestation of an underlying system and that caste and inequality make up the invisible structure. Race is the skin, and caste is the bones.

WHAT IS INSTITUTIONAL RACISM? HOW DO WE KNOW THAT IT STILL EXISTS?

Institutional racism refers to the work of institutions such as education, health care, and law enforcement to treat individuals differently based on race. The differential access and engagement with institutions is a major way that racism becomes real. Institutional racism highlights the ways that institutions, rather than individuals, drive racial inequalities. Some prefer terms such as "structural racism" or "systemic racism" to highlight how racism is part of the structure of society and a system that promotes inequality.

Some forms of institutional racism are explicit. For example, mortgage companies were clear about their practices of giving home mortgages to white families in certain areas while denying them to families of color with the same credentials and means. This practice, as noted earlier, was referred to as *redlining*.

Often, institutional racism is a silent code—whereby individuals and groups are treated differently. Predominantly Black schools lack the resources of predominantly white schools.[50] In chapter 4, we discuss how toxic waste dumps are often located closer to predominantly Black and brown than white areas. Policing is less effective in communities of color than in white communities. No one in America has ever viewed a video on television of a white person being choked to death by Black officers (the opposite of what happened to Eric Garner), or heard a report of Black officers breaking into the apartment of a white nurse and shooting her dead (as opposed to what happened to Breonna Taylor), or watching a white citizen having his neck crushed by a Black officer (such as the opposite of what happened to George Floyd). And on and on.

The proof of the existence of institutional racism is in the facts of differential treatment in almost all aspects of life. Black life expectancy, for example, is less than white life expectancy (see chapter 5). And this is not just a matter of having less money or lower socioeconomic class, although money and class are important. The data show that Blacks live shorter lives than whites at all educational levels. This is proof of systemic racism. One sign of the elimination of systemic racism is equality of life expectancy. The gap has closed slightly, and we look forward to it closing more and to all of us living our potential for as full a life as possible.

Chapter Four

WHY DO RACES DIFFER IN
DISEASE INCIDENCE?

In early March 2020, before the COVID-19 pandemic grabbed our attention, Joe was interviewed on Roland Martin's show, *Unfiltered*.[1] He stated that as soon as sufficient data became available on COVID-19 infection and death rates, it would show large disparities by socially defined race and ethnicity. We didn't need a crystal ball to foresee this, because racial inequalities have always existed for virtually every disease in our society. Moreover, during crises and hardships, poor and nonwhite communities are invariably hit the hardest.

Joe, of course, was right. The COVID-19 data were soon to show that 24 percent of those who died were African Americans, whereas African Americans made up 13 percent of the U.S. population.[2] On July 5, 2020, the *New York Times* reported that the virus positivity rates for African Americans and Latinos were 2.7 and 3.2 times greater than that for European Americans, respectively.[3] The rate of vaccination of African Americans and Latinos has fallen behind that of European Americans. In North Carolina, African Americans are 22 percent of the total population and 26 percent of the health care workforce but had received only 11 percent of vaccinations by January 30, 2021.[4] More recently, the CDC reports that, nationally, Blacks and Latinos still trail other ethnic groups in the rate of vaccination.[5]

As some of you might know, racial disparities in sickness and death have been a consistent feature of American society. You might have heard a variety of explanations for this, including the notion that if races differ genetically, it is only logical that they should differ in rates of sickness and death. This is a core premise of so-called racial medicine.[6] We place this idea under a microscope in this chapter and focus on the life-and-death consequences of living in a racialized society. We unpack the relationship between genetic variation and disease predisposition and demonstrate that any associations between genes and disease have nothing to do with race. On the contrary, the history and current reality of life in a racially stratified and racist society make people sick and shorten their lives.

In the United States, we routinely collect health data by race. Time after time, these data show glaring racial inequalities in disease and death. But the inequalities are not due to biological race; rather, they are due to the experience of living in a society that treats individuals differently based on their socially defined race. Racism has always been and is still a public health crisis.

MY DOCTOR'S OFFICE INTAKE FORM ASKS FOR MY AGE, SEX, AND RACE. WHY DOES IT ASK FOR MY RACE?

Socially defined races differ in their patterns of disease prevalence and rates of death. Thus, from the standpoint of current medical practice, it is reasonable for a physician to ask about a patient's socially defined race. The problem, of course, is that a patient's socially defined race is a poor proxy for understanding their predisposition for disease. Worse, there is still much confusion between social and biological conceptions of race in the medical community.

In May 2002, Sally Satel wrote a provocative article in the *New York Times* magazine titled "I Am a Racially Profiling Doctor." With that title, she was proudly playing off the notion of racial profiling that is a routine part of law enforcement. Satel made the point that it is appropriate and efficient to racially profile in health care because different races get different diseases and react differently to drugs and treatments. Unfortunately, Satel chalks up these differences to genetic factors, though often these differences do not hold up on closer inspection, and they differ over time and place.[7]

For example, one of the oldest myths is that Blacks have thicker skin and are less sensitive to pain, leading to their being undertreated for pain.[8] In this example, we see two mistakes: one about genetics and the other about the conjunction of race and genetics. First, there is a racial biological assumption that skin thickness and lower pain perception is due to genetics. Pain perception could be due to all sorts of factors and might have very little to do with skin thickness, and skin thickness might have very little to do with genetics. Second, even if genetics is somehow important in determining skin thickness and pain perception, we are aware of no data showing that these unknown genetic factors vary by frequency in different races. It is, in short, one racial assumption built on another and another.

A recent study conducted at the University of Virginia (UVA) Medical School found that false beliefs about racial differences in medically relevant traits persisted in first- to third-year medical students and residents (table 4.1).[9] For example, a third or more of first-year students thought that Blacks aged more slowly, their blood coagulated faster, and they had thicker skin and stronger immune systems than whites. Some of these misconceptions, such as the rate of aging and thicker skin, persist among the medical school's residents. These results are far from trivial. This is a premier medical school, so it is highly likely that the prevalence of these misconceptions is as great or greater among nurses and other health professionals and in less prestigious medical programs. Let's cease to make these conceptual errors.

The good news is that there has been movement toward reforming medical curricula to incorporate more nuanced understandings of human genetic variation and how such variation does not map onto socially defined race. We have authored several articles on this topic.[10] Many of our colleagues who understand the problems of racialized medicine work in medical schools, and others are giving talks at these schools. Alan has lectured at many medical schools and consulted with the University of Illinois Medical School. Joe has been active in the International Society for Evolution, Medicine, and Public Health (ISEMPH).[11] ISEMPH is a consortium of scientists and physicians dedicated to reforming medical instruction and practice so that it better incorporates evolutionary and anthropological perspectives, which are

TABLE 4.1
Some False Racial Beliefs Among Recent UVA Medical Students (percentages)

False Belief	1st Year	2nd Year	3rd Year	Residents
Blacks age more slowly than whites.	33.3	38.9	20.3	50.0
Black nerve endings are less sensitive than white.	12.7	19.4	0.0	14.3
Black blood coagulates more quickly than white.	46.0	23.6	5.1	14.3
Black skin is thicker than white.	63.5	58.3	37.3	100.0
Blacks have stronger immune system than whites.	33.3	20.8	5.1	14.3

Note: In general, medical students' false beliefs decline with years of training. The residents' pattern of false belief is not directly related to that for medical students, as many of these individuals were trained at other medical schools.

crucial in improving patient outcomes. These efforts are relatively new, and much work is still needed, but we are optimistic that instead of promoting racial biological thought, medical schools could become sites in support of antiracism.

WHEN DOCTORS, EPIDEMIOLOGISTS, AND OTHER MEDICAL SCIENTISTS SAY THAT RACE IS A RISK FACTOR, WHAT DO THEY MEAN?

A risk factor for disease simply means that having a condition or any "factor," such as being male, having a specific gene, living in a certain zip code, or having a bald spot, is statistically associated with an increased probability (or risk) of a disease. There is, or should be, no assumption of causality. However, with any risk factor, there is an association that might be worth exploring. Why do males have shorter life expectancies? The reasons—the processes and mechanisms—could be many.

For well over twenty years, we have contributed to a growing professional literature on the confusion in medical practice and biomedical research on the meaning of race as a risk factor. Just as not asking why baldness is associated with disease, not digging further to ask why race is a risk factor is not good science. It is giving up too soon. Why? Because not asking often leads to the assumption that the risk is due to genetics. And it certainly fails to take the next step to explore the reasons that race is a risk factor, those processes and mechanisms.

Socially defined race includes an individual's ancestry, position in society, and cultural practices. Thus, the raced and racialized experiences that people experience in our society result from these practices and much more. Due to the fact that socially defined races experience different social and environmental conditions, it is understandable that races might vary in disease rates and therefore that race would be a statistically valid risk factor in a variety of diseases, such as infectious (smallpox, tuberculosis, yellow fever, H1N1 influenza, HIV, COVID-19) or complex diseases such as cancer, diabetes, heart disease, and stroke.

Figures 4.1a and 4.1b show the ratio or relative risk of mortality (death) by age for Blacks compared with whites in 2015 for infectious diseases (influenza and septicemia) and complex, chronic diseases (heart disease and cancer). Septicemia is blood poisoning resulting from systematic bacterial infection. These graphs show that people who are socially defined as

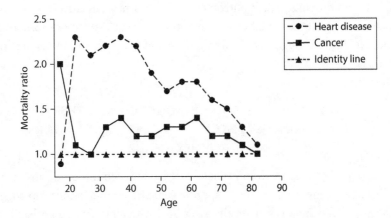

FIGURE 4.1A. The ratio of Black/white mortality by age in the United States in 2015 for heart disease and cancer, which are ranked as the number 1 and 2 sources of mortality in the United States. These diseases become the most significant contributors to mortality at later ages (so the youngest ages are not shown on this graph). From age twenty-two on, Blacks die from heart disease at a ratio of 2.3 - 1.1 times that of whites and from cancer at a ratio of 1.4 - 1.0 times that of whites. These ratios were more disproportionate in the twentieth century. See J. L. Graves, "Looking at the World through 'Race' Colored Glasses: The Influence of Ascertainment Bias on Biomedical Research and Practice," in *Mapping "Race": A Critical Reader on Health Disparities Research*, ed. Laura Gomez and Nancy Lopez (New Brunswick, NJ: Rutgers University Press, 2013). *Source*: Adapted and updated from Centers for Disease Control and Prevention, National Center for Health Statistics, National Vital Statistics System, "Leading Causes of Death 2015," https://www.cdc.gov/nchs/data/nvsr/nvsr66/nvsr66_05.pdf.

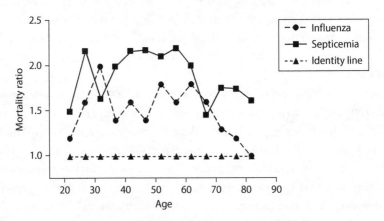

FIGURE 4.1B. The ratio of Black/white mortality by age in the United States in 2015 for influenza and septicemia, which rank in the top fifteen causes of death in the United States. They become the most significant contributors to mortality at later ages (so the youngest ages are not shown on this graph). From age twenty-two on, Blacks die from influenza at a ratio of 2.3:1.0 times that of whites, and from septicemia at a ratio of 2.17:1.46 times that of whites. These ratios were more disproportionate in the twentieth century. See J. L. Graves, "Looking at the World through 'Race' Colored Glasses: The Influence of Ascertainment Bias on Biomedical Research and Practice," in *Mapping "Race": A Critical Reader on Health Disparities Research*, ed. Laura Gomez and Nancy Lopez (New Brunswick, NJ: Rutgers University Press, 2013). *Source*: Centers for Disease Control and Prevention, National Center for Health Statistics, National Vital Statistics System, "Leading Causes of Death 2015," https://www.cdc.gov/nchs/nvss/index.htm.

Black in the United States display higher rates of death from these diseases across age. These ratios are also true for other leading causes of death, such as cerebrovascular disease, diabetes, chronic lower respiratory disease, HIV/AIDS, and, now, COVID-19. Thus, for virtually every kind of disease in the United States, socially defined race is and always has been a risk factor for developing and dying from that disease. It was this knowledge that led to Joe's prediction that COVID-19 data would also show significant disparity by socially defined race.

Yes, race is a risk factor. Almost every disease we are aware of varies by race. But that is just the first question. The more important questions to ask have to do with how consistent those results are and why disease and death rates vary by race.

ARE THERE RACE-SPECIFIC DISEASES?

No.

However, as we noted earlier, the prevalence of diseases and mortality rates differs in various socially defined racial groups. For example, infectious disease was a major source of death for Amerindian nations after the colonization of the Americas by Europeans. Smallpox first appeared in the Amerindian populations of Santo Domingo in 1518. It is likely that the English brought smallpox to the northeast by the early seventeenth century. As Snow and Lanphear note, John Josselyn wrote that by the 1630s, the Sagamore nation was reduced from 30,000 to about 300 by smallpox.[12] The high susceptibility of Amerindians to smallpox resulted from the fact that the virus was new to their populations, their means of subsistence had been disrupted because of colonialism, and the lower human leucocyte antigen (HLA) diversity of Amerindians compared with Africans, Europeans, and East Asians. The HLA system is encoded by the histocompatibility gene complex (MHC) and is responsible for the regulation of the immune response in all mammals. The lower HLA diversity of Amerindians resulted from the serial founder effects that established them in the Western Hemisphere and lowered their genetic diversity compared with the Eurasian populations from which they descended.

African Americans also suffered higher death rates compared with European Americans in the smallpox epidemic of 1862–65.[13] W. E. B. DuBois's sociological study, *The Philadelphia Negro*, published in 1899, also traced the disparate rates of tuberculosis to the poor social conditions experienced by African Americans in that city.[14]

Similarly, the disproportionate incidence of COVID-19 in African Americans and Latinos results from differential exposure to the virus, not increased genetic susceptibilities. The 2019 coronavirus is new to humans, most likely having originated in bats. Bioinformatic analysis of HLA loci that might confer resistance to or greater severity of COVID-19 found some HLA variants that were extremely rare and not differentiated by population.[15] Finally, in summer 2020, genome-wide association studies (GWAS) of Italian and Spanish patients identified variants that led to more severe cases.[16] Later that summer, it was determined that a segment of chromosome 3 associated with greater severity of COVID-19 most likely originated in Neanderthals (who once interbred with modern humans in Eurasia, not Africa).[17] African Americans can display these segments as well but at lower frequency, because the variants result from their European ancestry (derived from the rape of African women during slavery).

The prevalence of diseases with relatively simple genetic foundations such as sickle cell anemia, Tay-Sachs, and cystic fibrosis differs by population (but not race). For example, populations that live in high malaria transmission zones have elevated frequencies of the sickle cell (HbS) allele. Individuals who are heterozygous for this trait (HbA/HbS)—that is, have one copy of the sickle cell allele—display resistance to malaria; homozygous individuals without a copy of the sickle cell allele (HbA/HbA) display elevated death rates due to malaria; and individuals that are homozygous (HbS/HbS) with two copies of the sickle cell allele are at greater risk of death.

Sickle cell allele rates vary by region and population. Kenyans living at high altitude have very low frequencies of HbS, whereas low-altitude Kenyans have elevated frequencies of this allele. Greeks and Saudi Arabians have higher frequencies of HbS than native South Africans, as the former climates supported malaria transmission and the cape of southern Africa did not. The frequency of the HbS allele in African Americans in the United States is about 5 percent. This is lower than the frequency of the mutation (12–14 percent) in western and central African populations from which African Americans are descended, as malaria transmission in the United States dropped steadily over the nineteenth and twentieth centuries. Without malaria transmission, the HbS allele conferred no advantage and was therefore reduced in frequency by natural selection.

Earlier, we noted that the prevalence and mortality rates for complex diseases (e.g., heart and cancer) differ by socially defined race in the United States. A strong indication that these differences are not driven by genes is the fact that the rates of complex disease differ within groups of the same or similar ancestry by country of origin. Indeed, the rates of heart and other complex diseases in western Africa are considerably lower than those experienced by persons of African descent in the United States.

However, within a single generation, West African immigrants take on the disease pattern observed among African Americans. Prior to 1960, virtually no persons of African descent immigrated to the United States. This changed when biased immigration quotas were struck down by the Hart-Celler Immigration Act of 1965. Today, individuals of African descent are one of the largest groups entering the country. A series of studies have shown that although persons of African descent not born in the United States initially have a health advantage compared with African Americans, after living in the United States, these groups begin to take on the health profile of African Americans. These studies also showed that the health decline was greatest for immigrants from nations in which Europeans were the majority, suggesting that it resulted from their previous experience with racism.[18] Specifically, it is thought that these individuals experienced early-life adversity due to institutional racism in their countries of origin, making them even more vulnerable to the racialized health conditions in the United States. The role of early-life adversity is well established in accounting for disease in later life.[19]

DOES BLOOD DIFFER BY RACE? CAN I DONATE TO OR RECEIVE A BLOOD TRANSFUSION OR AN ORGAN FROM A PERSON OF ANOTHER RACE?

Former Hawaiian Senator Daniel Inouye was critically wounded in World War II. He was a member of the 100th Infantry, 442nd Regimental combat team. They were an all-Japanese-American unit, called the "Nisei soldiers," and many were recruited from internment camps. He received a life-saving blood transfusion from the Negro blood supply. At this time, the U.S. Army continued to mistakenly segregate blood supplies of whites and Blacks despite full awareness that anyone could receive blood from anyone else if their blood group genotype/phenotype matched, and that blood groups had nothing to do with race.

The ability of an individual to receive a blood transfusion (red blood cells) depends on variation at two major loci: blood group (A, B, O) and Rh factor (+ or −). Furthermore, variation at the Kell, Kidd, MNS, Duffy, and HLA (major histocompatibility) loci can influence blood group compatibility. Also, the capacity to successfully receive blood platelets depends on variation in the human platelet antigen system (HPA).

Human populations differ in their allele frequencies in all these systems. However, blood transfusion compatibility depends on the genotypes of the individuals at the key loci, not their socially defined race. Individual-level variation determines compatibility.

Given that genes are inherited in families, the best choice for a blood or tissue donor will always be a close relative. For example, if a man has children with two different women (one socially defined as white and the other as Black), it is more likely that the siblings will be good blood donor matches than unrelated individuals in either socially defined race.

The major human blood groups (A, B, O) were discovered by Karl Landsteiner in 1900. These groups are characterized by antigens (molecules such as proteins or sugars that elicit an immune response). The A and B antigens are sugars, and O is the absence of either the A or B antigen. The frequency of the ABO alleles differs in human populations. For example, recent surveys have found that the frequencies of blood type O are 70.9, 65.6, and 34.0 percent in African Americans, European Americans, and Chinese (in China), respectively. Furthermore, blood group compatibility requires the matching of at least both ABO genotype and Rh genotype. Determining the frequency of the two locus blood group phenotypes requires multiplying the frequencies of the individual loci (by the rules of independent assortment). The frequency of the Rh+ allele in these populations is 97 to 99 percent, 83 to 85 percent, and 93 to 99 percent, respectively. Thus, the frequency of the ORh+ phenotype in these groups would be 68.7–70.1, 54.0–55.7, and 31.6–33.7 percent, respectively. It follows that if ABO and Rh phenotype are the primary factors determining blood transfusion compatibility, a person of East Asian descent, such as Senator Inouye, with the ORh+ phenotype, would have a greater probability of getting the right blood transfusion from someone of African American descent in the United States than someone of East Asian descent.

For patients who receive multiple transfusions, genetic variation at the secondary (and tertiary) blood compatibility loci become important. These are the Kell, Kidd, MNS, Duffy, HPA, and HLA gene systems. As with the major blood compatibility loci, the frequencies of the gene encoding the antigens produced by these systems differ by population. For example, the Kell system has two major codominant alleles: K and k. Codominance means that both the K and k proteins (antigens) are made in persons with the Kk genotype. An antigen is any molecule (protein or carbohydrate) that elicits an immune response. The K antigen occurs at the lowest frequency in African Americans (0.02), is slightly higher in European Americans, and is highest in some Middle Eastern Arab populations (0.250). The K antigen, Rh phenotype frequencies are KRh−/kRh+ 0.91, 0.98; KRh+/KrRh− 0.0002, not detected; and KRh+/KRh+ 0.088, 0.02 in European Americans and African Americans, respectively.[20]

The take-home message here is that blood transfusion is determined not by socially defined race but by an individual's genotype at the primary compatibility loci A, B, O and Rh, as well as the secondary and tertiary loci. Individuals who carry the more common alleles at these genetic systems will find it easier to find a suitable blood donor. Individuals with rare alleles will find it more difficult to find a suitable donor. In all cases, close relatives will be the most likely suitable blood donors.

The ability to receive tissue donations is similar to blood donation. In this case, the major compatibility loci are the human leucocyte antigens (HLA) system. The HLA antigens are proteins found on the surface of virtually every cell in animals. The purpose of these proteins is to allow the recognition of foreign proteins and sugars from organisms (such as viruses, bacteria, and other parasites).

In humans, there are three major HLA loci (HLA-A, HLA-B, and HLA-C), and five class 2 loci (HLA-DPA1, HLA-DQA1, HLA-DQB1, HLA-DRA, and HLA-DRB1), along with some minor loci. At the HLA-A locus, 68 common alleles account for about 4,340 antigens, HLA-B, 125 common alleles account for 5,212 antigens; and HLA-C, 44 common alleles for 3,930 antigens. These staggering numbers explain how our HLA system protects us from pathogens but also why tissue donation is so difficult. The Be the Match Foundation requires that at least 6 of 8 HLA markers match to facilitate a successful bone marrow donation.[21] In some cases, more matches are required.

Just as in blood donation, tissue donation works best with a close relative (brother, sister, first cousins), who share more genes with an individual by descent and thus have a greater likelihood of sharing HLA genotypes. HLA allele frequency varies by population, and sub-Saharan Africans have the greatest diversity of HLA alleles. African Americans are less likely to find a suitable donor match for tissue transplantation due to the fact that low numbers of this group have registered for tissue donation and because of their greater HLA diversity. In some cases, individuals have received tissue donations across socially defined race.

A study published in 2004 found that the number of cadaveric renal transplants performed between 1996 and 2001, white-white, white-Black, Black-white, and Black-Black transplants accounted for 66, 23, 5, and 6 percent, respectively.[22] This meant that Black patients received 17 percent more renal transplants across socially defined race than from within socially defined race. With regard to the tissue match genetics, this makes sense, as Europeans have lower genetic diversity at HLA-A loci compared with Africans and because African Americans have on average 16 percent European genetic ancestry.

In conclusion, the idea that blood varies by race is a myth that we can put to rest. Race is not in the blood.

ARE CERTAIN MEDICINES MORE EFFECTIVE DEPENDING ON YOUR RACE? SHOULD A PILL BE COLOR-BLIND?

Yes, a pill should be color-blind. The efficiency and rate of metabolism of medicines might vary by genes, but that has nothing to do with social race.

The genes that are involved in an individual's drug metabolism are ancient. Many of them are conserved across vertebrate animals. Their primary purpose was to detoxify compounds associated with plant and fungal foods. However, ancestral mammals never encountered these natural sources of toxins at the concentration that we see in modern drugs. Thus, our body's lack of capacity to react to modern high-purity and -potency drugs is an example of an evolutionary mismatch. Evolutionary mismatches in humans often result from the fact that our technology and cultures have evolved at a much faster rate than our genomes. This is why in virtually every ad you watch on television for a medication, the small print and low-volume narration for its potential side effects might last the entire length of the commercial.

Genes associated with drug metabolism could vary in frequency by population and therefore can be a factor in how socially defined races respond to specific drugs. For example, in 2014, the attorney general for Hawaii, David Louie, filed a lawsuit against Bristol-Myers Squibb and Sanofi-Aventis, the manufacturers and distributors of the drug Plavix (brand name for clopidogrel).[23] In the suit, Louie alleged that the manufacturers engaged in unfair and deceptive marketing of the blood-thinning drug. Plavix is designed to prevent strokes and heart attacks. The suit claimed that the manufacturers did not disclose information related to how certain genetic traits alter the effectiveness of the drug and potentially lead to complications such as gastrointestinal bleeding. Studies of the drug found that 38–79 percent of Pacific Islanders and 40–50 percent of East Asians might respond poorly to Plavix.

The inability to successfully metabolize clopidogrel is associated with variation in the cytochrome P450 (CYP) enzymes. These enzymes are part of the cellular respiration chain. The CYP2C19*2 allele produces a defective protein, resulting in poor metabolism. The allele occurs in high frequency in East Asians, South Indians, and Pacific Islanders but at very low frequency in Africans and Europeans. Genetic variation at the CYP2D6 locus is particularly important because it enables a variety of drugs to be metabolized. There is also considerable genetic variation at this locus.[24]

To demonstrate that variation at this locus is not "racial" in the context of social definitions, the frequency of allele *1 is 83 percent in African Americans (of western and central African and European descent) but 27.8 percent in Tanzanians (East Africans) and 85.6 percent in Zimbabweans (southeast Africans). It is 27 percent in East Asians, 43 percent in Han Chinese, 43 percent in Japanese, and 0.49 in Koreans. Thus, this allele occurs both more and less frequently in Africans than Asians, depending on the African population.

Understanding how human genetic variation impacts drug metabolism reveals how ridiculous the notion of a "race-specific" medicine actually is. The drug BiDil was designed to be more effective than traditional medicines for congestive heart failure. BiDil is a combination pill that included an antioxidant and nitric oxide inhibitor. It was originally designed as a better therapeutic for anyone with heart disease. However, it became racialized when it was shown that although the drug didn't significantly improve heart disease outcomes for the overall population tested, it did so for the African American subsample in the study. The drug was then tested in the African American Heart Failure Trial (A-HeFT). As a result of the strong performance in that study, the manufacturer, NitroMed, sought and received permission from the FDA to market BiDil as the first "race-specific" drug. This right was granted despite the fact that BiDil was never extensively tested with other racial or ethnic groups. BiDil failed to capture a significant portion of the market for a number of reasons, including the fact that it was a self-limited market, its cost was high ($1,200–$2,800 per year), generics were available that could do the same job, and, finally, African American patients for the most part did not desire a "race-tailored" drug.[25]

The lesson of BiDil has changed the conversation in medicine from race as a proxy for biological, environmental, and social variation to the notion of personalized medicine. This sea change resulted in part from recognition that the key variables determining how a drug will metabolize in a specific patient are related to the individual's genome and their own environmental history and current situation. The advent of cheap whole-genome-sequencing technology is ushering in the day when this sequencing will be routinely offered to patients with sufficient financial resources, allowing for therapies tailored to a patient's specific genome and environmental exposures.

IN THE UNITED STATES, RACES DIFFER IN THE RATES OF THE MOST COMPLEX DISEASES, SUCH AS DIABETES, HIGH BLOOD PRESSURE, CONGESTIVE HEART FAILURE, AND CANCER. WHAT EXPLAINS THESE DIFFERENCES IF THE GENETIC DIFFERENCES ARE SO MINIMAL?

The short answer is that disease rates vary by race because growing up Black or brown takes a toll on the body. Both racism and being racialized cause health inequalities.

Now let's step back. An individual's disease risk is influenced by three sources:

1. genes,
2. the environment, and
3. chance.

The racial health disparity observed in modern societies is primarily driven by the first two causes, but not in ways that most people appreciate. The disparities we see in complex disease rates in the United States cannot be explained by allele frequency differences between socially defined racial groups. If we were to posit a simple genetic explanation of disease differential by socially defined race, we would expect persons of primarily European ancestry to bear a higher disease burden than those of African ancestry. This is because genome-wide studies show that Europeans actually carry a greater load of deleterious mutations than Africans.[26]

This is illustrated by examining risk alleles for hypertension and their frequency in African Americans and European Americans. Joe examined thirty-three such loci associated with hypertension in a prior paper and found that

African Americans actually had higher frequencies of the protective alleles at many loci.[27] Other studies have shown that when African Americans had higher frequencies of the risk allele, this was only true for persons of African descent in the United States. Persons of African descent with the risk allele living in western Africa showed no such elevation of risk.[28] Therefore, it is not genetic predisposition that accounts for African American health disparity; rather, it is American institutional racism that produces a toxic environmental effect on otherwise healthy genomes.

The reason that allele frequency differences do not readily tell us much about complex disease causality and risk is because of the complicated genomic foundations of these diseases. The more complicated the genomic foundation, the harder it is to compare populations, particularly those that significantly differ in linkage disequilibrium (LD) levels. Linkage disequilibrium is the nonrandom association between two alleles in a population due to the linkage of their loci along a segment of the chromosome. LD breaks down with time due to crossing-over events in meiosis (gamete formation). This means that older populations (sub-Saharan Africans) have smaller linkage blocks compared with newer populations such as Eurasians, Pacific Islanders, and Amerindians.

It's important to note that for this reason, a genetic variant that might be associated with a disease in one group might have no relationship at all with that disease in another group. For example, SNP rs9264942, which is associated with control of HIV, is located in the *HLB-C* gene in Europeans, but it has no effect on this trait in Africans. Although SNP rs2523608 is associated with the same trait in Africans, it is located in *HLB-B*.[29]

In addition, the polygenic character of disease predisposition means that these traits are strongly influenced by environment. Thus, in any genome-wide association study, before genetic causality can be assigned to any variant, it is crucial that the environment experienced by the organisms in question be equalized. This is always true in Joe's experimental work on life span and physiological performance in the fruit fly (*Drosophila melanogaster*) and bacterium *Escherichia coli*.[30]

These complexities and gene–environment interactions are prime reasons to take all claims of genetic causation of disease differential among any human populations—particularly socially defined races in the United States—with a heavy dose of skepticism. Racially subordinated populations (such as Amerindians, Latinx, and African Americans) have never lived in environments that are equivalent to the socially dominant European population.[31]

Table 3.1 presented the African American social experience as a single day. As emotionally traumatizing as these events were, we also know that they have had powerful effects on the health of persons who experienced them. A review of health literature between 2003 and 2013 showed that perceived racism impacted the health of African American women. This study found consistent evidence for the relationship between perceived racism and adverse birth outcomes (discussed later), illness incidence, and cancer and tumor risk. Inconsistent results were found for heart disease, and none of the studies could validate a relationship between high blood pressure and perceived discrimination.[32] These relationships are not unique to persons of African descent. It has also been documented for aboriginal Australians, who, though they share a dark skin phenotype with sub-Saharan Africans, are actually among the populations furthest in genetic distance from African Americans.[33] Racial discrimination is a form of social subordination, and social subordination has been shown to effect health status in a variety of animals (mammals and birds).[34]

Today, the mechanisms by which traumatic events affect physical and mental health are also better understood. One way that traumatic events (including racism) make their way "under the skin" is via epigenetic changes to the genome.[35] Epigenesis refers to non-nucleotide-based changes to the DNA molecule. This includes DNA methylation, short interfering RNAs (siRNA), micro RNAs (miRNA), and histone modification. DNA methylation refers to the addition of methyl ($-CH_3$) groups to DNA chain in places where a cytosine (C) is connected to a guanine (G) nucleotide. These modifications occur over an individual's life span. Epigenetic changes impact how genes are expressed and are known to play a substantial role in a variety of complex diseases such as heart, cerebrovascular, peripheral artery, liver, and inflammatory bowel disease, as well as cancer and a variety of mental illnesses.

Researchers examined more than two hundred fifty African American women and found a highly significant association between perceived racial discrimination and DNA methylation levels. In that study, the increased DNA methylation was associated with elevated levels of schizophrenia, bipolar disorder, and asthma.[36] There is evidence that epigenetic modifications play a major role in a variety of diseases that display racial disparities, including the big killers of heart disease and cancer.[37]

Finally, in chapter 1, we discussed how the environment had powerful effects on any organism's physical traits. The social subordination of racial

minorities has had significantly impacted the physical and biotic environments in which they have lived over the course of U.S. history. And in turn,
these physical and biotic environments have profound influences on the patterns of disease and mortality experienced by these minorities.

A glaring historical example of the importance of ecological conditions is
the dispossession of the Five Civilized Tribes (Cherokee, Chickasaw, Choctaw, Creek, and Seminole), who once occupied much of the modern states
of the Carolinas, Georgia, Kentucky, Tennessee, and Virginia during Andrew
Jackson's administration. The Indian Removal Act of 1830 was premised on
Jackson's deep racial mistrust of American Indians. This removal occurred
despite the fact that these tribes had agreed to change their way of life (many
adopting Christianity) to adhere to the cultural norms of white Americans.
The removal began with armed men seizing horses, cattle, and houses and
ejecting their occupants, assaulting any who dared to resist. Along the "trail
of tears," the tribes suffered from poor weather, inadequate food, disease,
depression (due to the loss of loved ones and homes), and mistreatment by
soldiers. The estimates of death from this forced removal of the Cherokees,
Choctaw, Creeks, and Seminoles was 4,000–5,000 (25 percent), 6,000 (15
percent), 500 (50 percent), and 500 (50 percent), respectively.[38]

This kind of environmental violence still occurs in the United States,
though perhaps not so blatantly. This is illustrated by the pattern of exposure
to the toxic wastes generated by our society. In 2007, it was estimated that
more than 9,222,000 people lived in neighborhoods within three kilometers of our nation's documented 413 commercial hazardous waste facilities.[39]
More than 5.1 million nonwhites—including 2.5 million Hispanics or Latinos, 1.8 million African Americans, 616,000 Asians/Pacific Islanders, and
62,000 American Indians—live in neighborhoods with one or more commercial hazardous waste facilities.

Host neighborhoods of commercial hazardous waste facilities are 56 percent nonwhite, whereas nonhost areas are only 30 percent nonwhite. The
percentages of African Americans, Hispanics/Latinos, and Asians/Pacific
Islanders in host neighborhoods are 1.7, 2.3, and 1.8 times greater (20 percent
vs. 12 percent, 27 percent vs. 12 percent, and 6.7 percent vs. 3.6 percent),
respectively, than their representation in the U.S. population.

Poverty rates in the host neighborhoods are 1.5 times greater than nonhost
areas (18 percent vs. 12 percent). More than 600,000 students in Massachusetts, New York, New Jersey, Michigan, and California attend nearly 1,200
public schools whose enrollment is largely made up of African American

and other ethnic minority children, and these schools are located within a half mile of federal Superfund or state-identified contaminated sites.[40]

More than 68 percent of African Americans live within thirty miles of a coal-fired power plant—the distance within which the maximum effects of the smokestack plume are expected to occur—compared with 56 percent of European Americans. Exposure to toxic materials such as heavy metals and pesticides has well-known impacts on complex diseases. It is possible to look up the exposure of various communities at the EPA's Toxic Release Inventory website.[41] One study found that in Southern California, census tracts with more than 15 percent Latino residences were exposed to 84.3 percent more toxic waste than those with less Latino residence. Similarly, census tracts with greater than 15 percent Asian residences were exposed to 33.7 percent more toxic waste. Finally, census tracts with more than 15 percent recipients with a bachelor's degree had 88.8 percent less exposure than those with less than a bachelor's degree.[42]

In conclusion, the answer to why health disparities for complex disease persist for racially subordinated communities in America can be found in the toxic social and physical environments that these communities have been forced to endure.

WHY ARE RESULTS FROM A BONE DENSITY TEST RACE-SPECIFIC?

It has long been thought that there are biologically based racial differences in bone density. Examples of this thinking go back to at least the 1800s. And it continues today. One day, the father of one of Alan's students, a practicing dentist, came to visit his osteology lab. The dentist saw a mandible (lower jawbone) and politely asked if he could pick it up. He then declared with confidence that the jaw, being robust and dense, must be from an African American male. It wasn't. But the myth of greater bone density of all African-derived individuals persists.

Possibly the most frequently cited study on race and bone density was published in a series of articles by biological anthropologist Mildred Trotter in the late 1950s and early 1960s.[43] Trotter and colleagues measured the bone densities of the long bones from cadavers that are still in the collections of the Cleveland Museum of Natural History. Information on age, sex, and race is available for the cadavers, as far as this demographic information could be ascertained. Most of the individuals were unclaimed bodies from the Cleveland morgue from the late 1800s and early 1900s. These individuals mainly seem to have lived around Cleveland in the mid- to late 1800s.

Trotter produced a number of graphs for different bones showing their densities on one axis and age on the other, with demarcations of individuals by sex (male or female) and race (white or Black). Her total sample was eighty individuals, with twenty in each of the four race/sex groups.

Her main conclusions were that (a) peak bone mass was greater in males than females and in Blacks than whites and (b) the rates of bone loss with age were rather comparable by race. Trotter's graphs contain regression lines for each race/sex group illustrating that bone density decreases in parallel fashion, at much the same rate, for each group.

What subsequent scientists seem to have taken from her study is the first part, that peak average densities vary by race and sex, and not the second part, that the rates of bone loss are similar for all groups.[44]

Because Alan has a condition that effects his bone turnover, about a decade ago, one of his doctors thought he should have his bone density tested. Alan traveled down Route 91 from western Massachusetts to Yale-New Haven Medical Center. There, the receptionist handed him a questionnaire to fill out with lots of the usual questions: name, address, sex, and age. He answered a bunch of additional questions about risk factors for low bone density, such as smoking, alcohol consumption, activity level, and diet. The form also asked him for his race.

He knew why they asked about his race. Simply, it is because race is considered a risk factor for osteoporosis or low bone density. But is it? What did this physician mean by "race"?

Alan has returned for a bone density test every summer. Each time, he fills out the questionnaire, and each time he asks the technician why they want to know his race. Answers range from a shrug of the shoulder to an explanation that it is needed so that the data can be interpreted, to an explanation that bone density varies by race. When he asks a follow-up question about why race is needed or why bone density varies by race, he sometimes gets an explanation that points to genetic differences. At this point, the technician usually gives him the same pamphlet on osteoporosis and risk factors that he always receives. Race is listed as a risk factor along with sex, age, genetics, and lifestyle. Is race different from genetics? Is it different from lifestyle?

So, what is under the hood of the machine that measures bone density? Bone densitometry, also called *dual-energy x-ray absorptiometry* (DEXA or

DXA), uses a very small dose of ionizing radiation to produce pictures, sort of like x-rays, typically focused on key bone density sites (spine and hips) to estimate bone density. One beam of x-rays is mainly absorbed by soft tissue and the other by bone, so from these, bone mineral density can be quantified. The technique is widely used throughout the world. It is simple, quick, and noninvasive.

So, why race? It has long been known that men have denser bones at a given age than women. Women are at increased risk of osteoporosis and bone fractures due to low density. The same is true of age: we lose bone with age. And the same is generally thought to be true of race.[45]

Alan gets his results. They include a graph with a dot that shows his bone density. On one axis, from the bottom of the page to the top, is bone mineral density from low to high; on the other axis, across the bottom, is age from 20 to 85. Three color zones run across the graph: green starts at the higher bone densities red at the lowest bone densities, and yellow in the middle. He finds a dot of his density, the point that crosses his age and bone density, right at the border of the green and yellow zones.

Now, the results are not purely about bone density; rather, they are "adjusted" based on comparing his results with those of other individuals of his age, sex, and race. It makes sense to compare him with others like him: apples to apples. It would not make sense to compare a twenty-year-old with an eighty-year-old or a male with a female.

But race? The calculation is based on an algorithm called *FRAX*. FRAX is widely available, and one can put one's own results into FRAX or even make up results and enter them into FRAX. Alan did both. It turns out that if he had said he was African American, his risk of osteoporosis would be seen as much greater, because the comparative database for Blacks has higher bone densities.

The use of race here as a risk factor is illustrative of a common set of mistakes, actually two leaps of faith: that racial differences in osteoporosis risk are genetic and that genetic differences break along racial lines.

Certainly, there are some genetic factors that could lead to greater or lesser peak bone mass and the rate of bone loss from that mass. But, of course, there are a host of other factors, like the ones Joe was asked about, including exercise and diet, that can affect a person's bone health.

In the end, what does race mean? Is it standing in for genetics? If so, we know that this is a deep problem. If genetics is the issue, why not ask instead about family history or collect some actual genetic information?

Let's also be clear that FRAX and osteoporosis is not the only example of how race gets baked into the science. Race is a demographic factor that is all too frequently used to "adjust" for risk. Lundy Braun of Brown University has written about how the spirometer has separate adjustment points for race.[46] Also, the interpretation of iron deficiency takes race into account, with the highly likely result of undercounting African American iron deficiency.[47] Finally, it has recently come to light that the diagnosis of a concussion is based on separate standards for African Americans than white football players.[48] Race is in the algorithms. It is baked into biomedicine.

In conclusion, we think we know why you are being asked your race, and that is because race is assumed to be a biologically stable risk factor for osteoporosis. But we see no reason that this should be the case despite how it is embedded in FRAX. So, the answer is bad medicine.

We hate to call out an entire industry and profession for making the conceptual error in thinking that race is a useful genetic construct. So, here's a simple challenge: prove to us that asking one's race helps to understand, diagnose, or treat bone loss. In particular, show us that this is the case after you have corrected for easily obtainable (and repeatable) information such as diet, exercise, and known genetic risk factors.[49]

Chapter Five

LIFE HISTORY, AGING, AND MORTALITY

Death and taxes. Both are guaranteed to happen. The former is guaranteed for everyone and all living things. It is the last of what biologists refer to as "life history events." The other important events include birth, maturation, menarche and menopause (the beginning and end of the ability to give birth), and aging (or *senescence*). In this chapter, we focus on myths about life history differences among races.

Evolutionary fitness is defined as "differential reproductive success." For organisms like ourselves, this involves how long we live and how much we reproduce. We calculate evolutionary fitness as the product of the probability of survival at a given age (l_x) times the average reproduction of individuals at that age (m_x). For example, if 100,000 babies are born in a given year, some fraction of them will not survive the first year. Similarly, of those who survive to age one, some fraction will not survive to their second birthday. We can continue following that group until every one of them has died (for humans, usually about one hundred years; see fig. 5.1).

Similarly, the average reproduction of the cohort is calculated by the number of live births to the females in that group each year. For humans, reproduction generally doesn't begin until females are around age thirteen, but this continues, with peak reproduction occurring in the twenties and declining until no reproduction occurs sometime in the fifties. Life histories are important because variation in the timing of life events might greatly shift fitness.

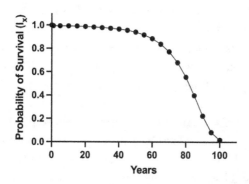

FIGURE 5.1. The age-specific probability of survival for Americans is plotted. The probability of survival drops steadily until all individuals of the original birth cohort (100,000 individuals) have died. Data from National Center for Vital Statistics, 2008.

Traits associated with life history include age of reproductive maturity, age-specific reproduction, survival rate, aging, and number and size of offspring at birth. The study of life histories in a variety of organisms (animals, plants, fungi) demonstrates that many life history traits are in opposition to one other, an adaptive trade-off. For example, life span and reproductive effort generally trade off against each other. In other words, traits and environmental conditions that tend to increase average reproduction generally decrease health and life span. We now know that these trade-offs play a crucial role in the health profiles of organisms.[1] Thus, at different points throughout a person's life, increased investment in reproductive function by either physiology or behavior affects the risk of many diseases and therefore life span.

Genetic factors contribute to interindividual and interpopulation variation in reproductive traits; however, the dominant source of variability is how the environment influences physical traits (phenotypic plasticity) during development and adult life. In our species, reproductive traits in both sexes evolved sensitivity to ecological conditions. This is reflected in contemporary associations of hormone concentrations with geographical setting, nutritional status, and physical activity level. Also, lifetime exposure to increased concentrations of sex hormones is associated with the risk of some cancers. As well, faster sexual maturation and higher number of births per age interval (parity) increases one's risk of diabetes and cardiovascular disease.[2]

To illustrate this compromise, one recent study examined progesterone (one of the reproductive hormones) levels in reproductively healthy females from different populations (Democratic Republic of the Congo, Bolivia, Nepal, Poland, and the United States). Women in the Congo are generally Africans; in Bolivia, Amerindian/European; in Nepal, Eurasian; and in Poland, European. The race/ethnic identity of women from the United States was not identified. This study showed that women from the Congo, Bolivia, Nepal, and Poland had statistically significantly lower levels of progesterone (200–250 picomoles/liter compared with 350 picomoles/liter) than American women. The authors suggest that this is due to the higher physical workload and lower energy balance of the non-American women. Thus, although genetic variation exists between these different groups of women, it was not a primary factor in determining their progesterone levels. In turn, the life history trade-off theory would predict that all other factors being equal, the non-American women would display lower average reproduction and longer average life span compared with American women. However, we know that in the countries represented in this study, all other factors are not equal.

WHY DO BLACKS SEEM TO AGE SLOWER THAN WHITES? (AND IS THAT TRUE?)

No. It is not true that Blacks age slower than whites. Indeed, the evidence we reviewed concerning sickness and death rates of African Americans in the previous chapter speaks volumes against this mistaken idea. In 1999, the life expectancy for whites was ~77 years compared to only 72 years for blacks. From 1999 to 2013, the life expectancy increased for both groups but the gap still remained. Thus, the data clearly show that longevity for Blacks has historically been lower than for whites in the United States.

This myth of slower aging of African Americans most likely tracks to the popular expression, "Black don't crack," which is associated with African American skin. It is true that darker skin sustains less ultraviolet damage than lighter skin. This means that as darker-complexed individuals age, they show less wrinkling in their skin, giving the appearance of younger age. This, of course, will be true of all dark-complexed individuals whose ancestry can be traced to any of the populations of tropical equatorial regions (Africa, Middle East, India).

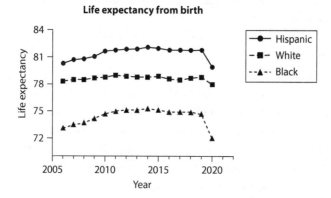

FIGURE 5.2. Life expectancy at birth from 2006 to 2020. The life expectancy for all socially defined racial groups increased and then declined in that period. Hispanics had the highest life expectancy, followed by white, with Blacks displaying the lowest. The drop for all groups in 2020 was due to COVID 2019. Blacks showed the steepest decline due to the virus.

However, to truly understand that this idea is false requires a better understanding of what aging actually is. Surprisingly, most people (including medical doctors and biomedical scientists) do not fully understand aging.[3] The evolutionary theory of aging has been validated in a large number of experiments in model organisms such as bacteria, round worms, and fruit flies.[4]

The first misconception that people often make is to associate aging with chronological time. If this assumption held, we would have a really difficult time explaining the life span of organisms with cellular structures similar to our own whose maximum life spans vary over orders of magnitude (May flies, one day; mice, three years; humans and lake sturgeon, 100-plus years; redwoods, more than 300 years; fungal fairy rings, 3,000 or more years; poplar clones, more than 10,000 years). In humans and organisms like us, it is better to think of our life spans as punctuated by stages: development, adolescence, adulthood, senescence (aging), and late life. Senescence does not begin until our net future expected reproduction has dropped to zero. At that point, the power of natural selection to correct the expression of genetic systems that might reduce our evolutionary fitness (the product of our age-specific survival and reproduction) consistently declines. This results in an increasing dysregulation of our molecular, cellular, and physiological function, eventually ending in death.

The inability of natural selection to address such late-life genetical disasters allowed our genome to accumulate alleles that contribute to aging via two mechanisms, mutation accumulation and antagonistic pleiotropy. Alleles governed by mutation accumulation have no effect during development, adolescence, and adulthood but negative effects during aging. Those that are governed by antagonistic pleiotropy are beneficial during the first three stages of life but detrimental during senescence (see table 5.1). Genes governed by antagonistic pleiotropy would be found in all world populations (although different variants and frequencies); those determined by mutation accumulation might be specific to certain populations. These genes would contribute to a variety of pathologies that begin to accelerate at later ages, including those impacting the leading sources of mortality in Western nations, such as cancer and heart disease.

However, aging is a complex phenomenon, with contributions from virtually every gene in the human genome, and therefore is strongly influenced by environmental factors. The best studies of the genetic contributions to life span find a maximum contribution of about 30 percent from genes and 70 percent from environment. In this chapter, we have already discussed the differential environments experienced by persons due to socially defined race. In this regard, studies of mortality rates with age have shown consistently that African Americans die sooner than European Americans and Asian Americans.[5] Insight into this pattern has been provided by new genomic methods that have been developed to measure biological (as opposed to chronological) age. Recent studies have found that DNA methylation can be used as a

TABLE 5.1
Population Genetic Mechanisms That Account for Aging

Mechanism	Early life	Late life	Examples
Mutation/selection balance	–	NA	Genes, such as progeria, do not contribute to aging; individuals rarely live to reproduce; frequency will vary in populations by mutation rate.
Mutation accumulation	0	–	Alzheimer's disease, Huntington's chorea; have no negative effects on early life; frequency in populations will be determined by genetic drift.
Antagonistic pleiotropy	+	–	Cancer genes play a beneficial role in development and maintenance of tissues. All populations will have these genes, although the variants might be different in specific groups.

Note: Mutation/selection balance is not an aging mechanism and doesn't really affect average life span (these alleles will be rare). Mutation accumulation effects will vary by population and could contribute to variation in rates of aging. Antagonistic pleiotropy governs genes that had beneficial effects early in life. These genes would be present in all humans (and in many related species) as well.

proxy for biological aging. We know that DNA methylation can result from differential stress and exposure to toxic environments.

One recent study found that in genes associated with age-specific cellular pathways, African Americans showed 4,930 methylation sites compared with only 469 in European Americans. The authors calculated that the increase in biological aging was gender-specific, with an increase of biological age of 2.04 years in a sample of middle-aged African American compared with age matched European American men.[6] Results from these kinds of studies tell us that the "Black does not crack" metaphor is just skin deep. Data on age-specific sickness and death in the United States explains that although Joe's skin might look good for someone his age, he is far more likely to die sooner than Alan or any other white male with a comparable occupation and social status.

Another general measure of aging (or weathering) is telomere length. Telomeres are the ends of chromosomes and are made up of repeated nucleotide sequences that act to protect chromosomes from deterioration or from fusion with other chromosomes. Telomeres shorten with age, but their rate of shortening varies and is accelerated by psychological stress. Shortened telomeres are associated with cellular health and aging. A recent community that engaged in a collaborative study in Tallahassee, Florida, found that self-reported unfair treatment attributed to race (or racial discrimination) is associated with shorter telomere lengths.[7]

The take-home message is that institutional racism still kills racially subordinated groups faster than those of the socially dominant populations. Racism, not biological race, increases aging.

WHY ARE THERE MORE BLACK THAN WHITE BABIES WITH LOW BIRTH WEIGHT?

Infant birth weights and infant mortality tell another vital story about how racism, not biological race, persists.

First, we present some facts and definitions. The average birth weight in the United States is around 7.5 pounds (3,400 grams). Birth weights are a function of the rate of growth and amount of time in utero (the longer or closer to full term, the higher the weight). Early birth is defined as *prematurity*, and being born small for the length of pregnancy is called *intrauterine growth retardation*.

The average birth weight of a Black baby in the United States in 2016–18 was about a pound (454 grams) less than that of a white baby. And the rate of low-birth-weight babies born to Black women (less than 2,500 grams) is nearly twice that of babies born to white women. These inequalities have persisted pretty much unchanged since the United States began to systematically collect birth weight and infant mortality data.

Infant birth weights are of great importance because they are the clearest risk factor for infant mortality (death before the first birthday) and future developmental problems. Today, Black babies die at a rate over two times that of white babies.

Study after study has shown that in the United States, the birth weights of infants of Black women are lower than those of infants of white women.[8] The extent to which the lower birth weights gap is due to life circumstances or genetic factors is unclear. However, given the consistency of the difference over time, regions of the country, and social strata and economic classes— combined with a bias toward genetic explanations— physicians and biomedical researchers have all too often favored a racial-genetic explanation.

In one study, the birth weight records of more than 90,000 infants from 1980 to 1995 from Illinois were used to test the importance of lived experience versus race.[9] Most of the infants were born to white (n = 44,046) and Black (n = 43,222) mothers who were themselves born in the United States. As in prior studies, the investigators found a clear difference in birth weights, such that babies born to the white women had an average weight of 7.58 pounds compared with 6.80 pounds for babies of Black women.

Notice this study specified the mothers' place of birth. That is because the authors wanted to test whether growing up in the United States, a racist country, with stresses from racism, might be a cause of the lower birth weights. Thus, the authors of this study, neonatologists Richard David and James Collins, added to their study the birth weight records of 3,135 infants of African-born Black women who had given birth in Chicago but spent at least part of their lives in Africa.

Because of the history of white men having babies with Black slaves, African Americans are genetically part white (we discuss this in more detail in chapter 9), whereas this is not the case with African women. Therefore, if racial genetics is the predominant explanation for the birth weight difference, one would expect that the babies of "purer" African women, those born in Africa and with less European admixture, would have the lowest birth weights.

However, the study found that the babies of Black women who were born in Africa had an average birth weight of 7.33 pounds, or just a quarter pound less than the babies born to white women. The close tracking of the distributions of babies born to white women and those of Black women who grew up in Africa is strikingly clear, as is the lag in birth weights of babies born to Black women who were born in the United States (figure 5.3). David and Collins conclude, "The birth-weight patterns of infants of African-born black women and U.S.-born white women are more closely related to one another than to the birth weights of infants of U.S.-born black women." Moreover, the results strongly—and dramatically—smash a genetic hypothesis and support a hypothesis that growing up under conditions of racism takes a lifelong toll.

What, then, explains these differences? The results we suggest strongly point to a lifelong consequences of living in a racist society. This has been called "weathering" by Geronimus.[10] Others have called this the developmental origins of health and disease (DoHaD).[11] These hypotheses are similar in that they consider how early life stresses can have lasting effects on life events, aging, and adult disease. These impacts have been shown in other mammals and birds, not just humans. Thus, we now know that the

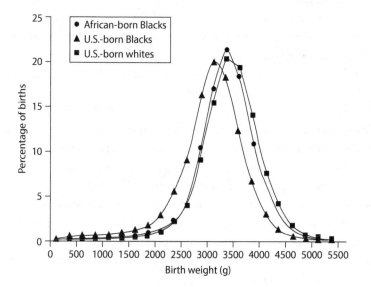

FIGURE 5.3. The distribution of birth weights for babies born to African-born Black women is very similar to that of babies born to U.S.-born whites. The birth weight distribution for both groups is higher than that of babies born to U.S.-born Black women. These results strongly suggest that African ancestry is not the primary factor contributing to the birth weight differential of U.S.-born Blacks and whites. *Source*: Richard David and James Collins, "Differing Birthweight Among Infants of U.S. Born Blacks, African-Born Blacks, and U.S. Born Whites," *New England Journal of Medicine* 337, no. 17 (1997): 1209-14

weathering of moms (and the epigenetic mechanisms of institutional racism) affects their babies' development in utero and forever.

DOES LIFE EXPECTANCY VARY BY RACE?

Yes. And this is one of the main areas in which we see the effects of persistent racism.

First, the overall average life expectancy from birth in the United States is shockingly low compared with other countries. In 2015, the life expectancy was 78.8 years. That was actually a slight decline from the prior year and a long-term trend of increasing life expectancy. Researchers point to problems like death from opioid addiction.[12] The United States currently ranks around forty-sixth in the world for life expectancy, almost six years below Japan and about four to five years below most countries in the European Union. It is almost four years below Canada as well as Cuba, Costa Rica, and Chile. Yet, we spend almost twice as much as any country on health care.

In addition to the overall poor record in health, there is a huge disparity in life expectancy by race and socioeconomic status. African American life expectancy has typically been about 5 to 10 percent shorter than that of whites. Recent data suggest some closing of the gap—to about four years—but much of that is due to increased deaths from addiction among whites. However, the differential death toll on Blacks due to COVID-19 might further widened the gap beyond four years (figure 5.1). These are real minutes, hours, days, months, and years of real lives lost.

Figure 5.4 shows the impact of race, gender, and educational attainment on life expectancy in the United States. First, note that Hispanic females and males have relatively high life expectancies. This is an example of what has been called the "Hispanic paradox." Hispanics generally have better health than one might expect given their position in U.S. society. They might be protected from more virulent forms of stress. And their being recent immigrants might insulate them from the weathering effects of U.S. racism.

Education is a huge predictor of life expectancy. For all groups, increasing education from less than a high school degree to a college degree increases life expectancy by around a decade. Stay in school!

Finally, Blacks have lower life expectancies than whites and Hispanics at all levels of education. For example, Black males without a high school degree have life expectancies in the mid-sixties. Even with a college degree,

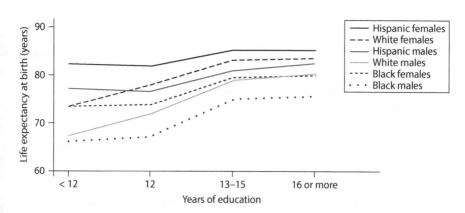

FIGURE 5.4. Life expectancy by race and years of education. Life expectancy increases with years of education for all groups, but the gap between that for whites and Blacks remains. Hispanic males and females show greater life expectancy than either whites or Blacks. *Source*: S. J. Olshansky, T. Antonucci, L. Berkman, R. H. Binstock, A. Boersch-Supan, J. T. Cacioppo, and J. Rowe, "Differences in Life Expectancy Due to Race and Educational Differences Are Widening, and Many May Not Catch Up," *Health Affairs*, 31, no. 8 (2012): 1803-13.

their life expectancy is less than the average white life expectancy and below that of a Hispanic male without a high school degree.

Yes, life expectancy varies by race. This is one of the most important indicators of racial inequality.

IF RACES ARE NOT BIOLOGICAL, WHY ARE THERE SUCH PERSISTENT DIFFERENCES IN HEALTH AMONG RACES?

The simple answer to this question is that the social and physical aspects of the environment that contribute to health disparities have not changed enough to eliminate the disparities. This reality has been brought into sharper focus by the COVID-19 pandemic. As this virus was new to our species, no population had genetically based resistance to it, in the way that endemic diseases such as malaria (sporozoan parasite), yellow fever (flavivirus), or sleeping sickness (trypanosomiasis) do. The rates of infection and death from COVID-19 that we are now experiencing are explained by differential exposure and lack of adequate health care. Indeed, the rates we see in Latinx populations is also associated with a fear among undocumented people to seek medical treatment because they might be deported.

Thus, the differential patterns of sickness and death that we see in the United States is one of the most powerful signs of institutional racism. The killings of George Floyd, Breonna Taylor, and Ahmaud Aubrey are just the tip of the iceberg of this murderous system. These stark ongoing disparities are a call for all people of goodwill to take immediate action.

CONCLUSIONS

Humans have a specific life history marked by a long period of early development and slow maturation. However, there is no evidence that this pattern varies by continent. Rather, there is strong evidence that life histories vary depending on the stresses of different environments. Adaptations to early stresses can have impacts on later events such as birth weight and telomere length. These results show that racism, once it gets into the body, can have long-lasting effects.

Chapter Six

ATHLETICS, BODIES, AND ABILITIES

How do we think about the connection between athletic ability and intelligence? Are they connected in any way? Do individuals who are smart tend to be more athletic and physically able, or is the reverse more likely? And what about races? Are some races more athletic than others? Are members of some races more intelligent?

Up to the 1930s, the general idea was that there was a connection between athletic ability and intelligence. Whites from elite schools such as Harvard, Yale, and Princeton were viewed as smart and intelligent as well as handsome and athletic. They played football and other sports. The Kennedy clan, Boston Brahmins and mainly Harvard educated, were often photographed and filmed playing football and sailing. President John Kennedy was handsome, trim, and smart.

After the Age of Discovery, individuals of European decent thought that they were more intelligent than those of African and Asian descent. That's perhaps the oldest and most central myth of race. And until the 1930s, they also thought that they were the most athletic, strongest, and most virile of the races. Whites dominated the Olympics and professional sports. Of course, few non-European countries fielded a full-fledged and well-trained Olympics team, but that did not seem to impact the view of European athletic superiority. And in the United States, all of the professional sports leagues barred nonwhites. It is easy to dominate when you have no competition.

There were always counter facts to European athletic superiority. Heavy-weight boxer Jack Johnson (1878–1946) was a champion and without doubt the best boxer in the age of Black terror (the Jim Crow era). Satchel Paige (1906–82) started playing baseball in the Negro Leagues in the 1920s and might well have been the best pitcher of all time.

The myth of European athletic superiority took a tumble in the 1930s. In 1936, African American Jesse Owens won four gold medals at the Berlin Olym-pics, to the dismay of Adolf Hitler. Two years later, African American heavy-weight boxer Joe Lewis knocked out Max Schmeling, the German champion, to recapture the world heavyweight boxing crown. Boxing required feats of speed, strength, and cunning; thus it was no longer possible to hide the fact that non-Europeans might be the athletic equals of Europeans.

Instead of an acknowledgment that excellent athletic performance might be found in individuals of all socially defined races, another myth emerged: that there is an inverse relationship between athletic and intellectual ability. And it was racialized. Whites are smart and Black and brown folks are ath-letic. Here, we unpack this myth.

DO RACES DIFFER IN THEIR ATHLETIC ABILITY?

No. Individuals certainly do vary in athletic abilities. There also are data suggesting that genetic variation among groups might play a role in group differences in specific athletic feats. But that variation is not racial variation.

In prior chapters, we established that human beings cannot be classified into biological races. So, we could stop right here. The question of different abilities of any sort by biological race makes no sense, because biological races make no sense. They do not exist.

But clearly, many individuals today wonder about the dominance of African Americans in sports such as basketball and whether that could be explained by genetic differences among groups of people. A common ques-tion we hear in private conversations is some version of "If race is not genet-ics, then why are there so many excellent Black athletes?" So, let's explore these ideas head on!

First, to answer the question about the reality and likely roots of any pur-ported racial differences in athletic ability, we have to be clear about what is meant by "*athletic ability*" and how it is measured. A general definition is "the strength, co-ordination, speed of reflexes and other qualities and skills which are required for a person to excel in a sport." By that definition, many sports require much the same things: coordination the same muscle groups, as well as stamina, to be successful. However, we quickly see how the definition gets racially subverted.

The 2008 Summer Olympics in Beijing saw stellar performances by Usain Bolt (Jamaica, primarily of African descent) and Michael Phelps (United States, European descent), but only Bolt's athletic accomplishment was racialized.[1] Phelps's success was discussed in the context of the sacrifices made by his single mother and the outstanding coaching he received throughout his career. His success was due to circumstances and perseverance. Bolt's success, on the other hand, was discussed in the context of genomic variants associated with superior Jamaican sprinting ability.

What is clear is that athletic performance, just like aging or intelligence, however defined, is a complex physical and mental trait involving a set of interacting genes of interacting genes, environment, culture, and opportunities. For example, a 2006 study of the genetics of athletic performance found 165 autosomal, 5 X-linked, and 17 mitochondrial loci that had some relationship to performance. A year later, the same research group found 214 autosomal, 7 X-linked, and 18 mitochondrial loci that related to athletic performance.[2] These genetic markers were found by comparing elite athletes and nonathletes using candidate gene methods, a standard method at that time.

After reviewing these studies, in 2013, Joe found that there was no basis for differentiating socially defined racial groups according to genes associated with athletic performance.[3] Several more recent studies have failed to find any racial component to the genomic foundation of athletic performance. For example, a review of the literature from 1997 to 2014 found 120 genetic markers (77 endurance, 43 power/strength) associated with elite athlete status. A genome-wide association study (GWAS) found 6 markers associated with athleticism in a sample of African-American, Jamaican, Japanese, and Russian elite athletes.[4] The fact that these markers were found in persons of widely divergent descent also shows that there is no racial basis for athletic performance.

Furthermore, these genetic variants are pretty evenly distributed across continents. For example, the F_{ST} value for one variant, the nuclear factor 1 transcription factor SNP, rs1572312, is only 0.065. (Remember that an F_{ST} below 0.25 suggests that there is little variation among groups.) Indeed, for the seventy-six populations in the ALFRED (allele frequency) database, the frequencies of the allele are virtually the same in all populations. Also, the F_{ST} value for the SNP associated with muscle power (rs1815739) in the *ACTN3* (alpha-actinin 3) gene that is expressed in type II muscle fibers is 0.139. The range of this SNP in Africa is 1.00–0.65; in Europe, 0.760–0.500; and in East Asia, 0.650–0.500. This means that in virtually every country measured, at least half the population carries the SNP for increased muscle power.[5]

Thus, the science of human genetic variation does not support the idea that any large (continental) human groups are necessarily more athletic overall than others. Elite athletes' ability is the complex result of genomic variants, gene expression (including epigenetic effects), and metabolic efficiency, as well as access to training facilities and coaching.[6]

Biomechanical differences also influence the capacity of specific populations to produce world-class athletes. For example, although ancestry associated with high-altitude adaptation is a definite advantage for endurance events, due to the need to maximize oxygen metabolism, excellence in long-distance running is more likely to some groups such as Ethiopians as opposed to others, such as Himalayans or Andeans. This is because other biometric attributes, such as body mass and proportion, favor Ethiopians.

Culture also plays a significant role in which sports athletes from different societies engage in. The success of Ethiopians and Kenyans in long-distance running is directly related to the lack of transportation available in the communities that have spawned the recent wave of world-class runners from those nations.[7]

Similarly, the success of male and female Jamaican sprinters can be directly related to the cultural influences that funnel athletes into this sport.[8] Of the men who have completed the 100-meter dash in under 9.8 seconds, six are from Jamaica and three are African Americans. There are 41 million African Americans and only 2.8 million Jamaicans. For example, since 1910, Jamaica's premier athletic event has been the boy's high school track and field championship. Success breeds success.

Another example of how opportunity and culture influence sport is the way in which basketball transitioned from a sport dominated by various European ethnic groups to an inner-city game dominated by African Americans. In the early part of the twentieth century, basketball was dominated by Jewish, Irish, and Italian teams before it became "integrated."

Now, the complexion of basketball is changing again, rising in popularity in Eastern Europe and throughout the world. As a result, some of the brightest stars of today's NBA, such as Luka Dončić from Slovenia and Rudy Gobert from France, come from Europe. During the 2019–20 NBA season, 108 foreign-born players were on NBA rosters. Europeans now outnumber

European-Americans in the NBA. Is that because Europeans are more natural basketball players? This is highly unlikely. Rather, we believe that most European-Americans have shifted away from basketball while Europeans are embracing the sport.

ARE FINNS THE WORLD'S BEST ATHLETES?

Finns? If you were envisioning sprints, I bet you thought we were going to say Jamaicans, or if you are a fan of marathons, you might be inclined to answer "Ethiopians." Or Americans if you are into total Olympic medal counts. Finns? Let's get to it!

How do we evaluate extraordinary athletic ability? We could examine direct test strength, speed, and jumping ability, but those data are not available for random samples. Another way to think about athletic ability is via Olympic medal counts—flaws, biases (which we will discuss), and all.

The highest Finland has ever ranked in the medal count in the Winter Olympics is second, in 1924.[9] Since 1992, the medal count for Finland has placed its athletes between eighth and twenty-fourth. The pattern for the Summer Olympics is similar. In 1924, when Finns ranked second in total medal count, the Finnish team was led by Paavo Nurmi, one of the greatest distance runners of all time. Nurmi won gold in the 1,500- and 5,000-meter, individual cross-country, 3,000-meter team, and men's cross-country team races. If we asked you whether Finns were the world's best athletes in 1928, what do you suppose your answer might be?

Since 1992, the Finnish medal totals have been between twenty-ninth and seventy-eighth highest. In the Winter Olympics, the Finns' best sport has been cross-country skiing, which makes sense given this country's climate, with 80 medals earned. In the Summer Olympics, the best performance has been in track and field, with a total of 114 medals. Only 16 nations participated in the 1924 Winter Olympics compared with 92 in 2018. The 2016 Summer Olympics had 205 nations competing.

Based on everything we know about athletic performance as a complex trait, we would never expect a country with a population as small as Finland's to produce the world's best athletes. The population in 1924 was only 3,272,000, and in 2020 it was 5,553,000. Finland was settled by founder populations from northern Europe, so it has less genetic diversity than the rest of Europe. For example, its frequency of the genetic diseases common to northern Europeans is considerably lower. For example, the frequency of cystic fibrosis in Finland is a tenth that of its neighbors.[10] It is possible that some adaptation unique to Finland occurred, but that is highly unlikely. And, too, countries are not populations.

Athletic ability as a quantitative trait is distributed in a normal (bell curve) shape. Using the tools of classic genetics, we can calculate the number of physical classes of quantitative traits. Applying the formula for physical classes ($2N + 1$) using the 239 loci that have been associated with athletic ability results in 479 categories of athletic ability. This means that 239 classes fall above the central point (mean, median, and mode constitute a normal distribution) and 239 below. Elite athletes would be individuals who fall into the highest physical classes. The size of a population and its genetic diversity might be factors in producing people who fall into the extreme high ends of the athletic distribution, again, where we would expect to find people who are elite athletes.

The NCAA listed about 35,000 female and male basketball players. If we assumed there were fifty sports in the United States, all with 35,000 athletes (a generous assumption), and calculated them as a fraction of the age classes of where we observe major sports activity (18–44), less than 1.4 percent of Americans would be considered elite athletes.

One way of determining the performance of a nation's elite athletes is the number of medals earned in the Olympics. In 2007, Joe ran a statistical analysis of total Olympic medal count and showed that a nation's population size and gross domestic product were highly significant variables in determining how many medals its athletes earned. Nations such as the United States, China, France, Germany, Japan, Russia, and the United Kingdom dominate the Olympics because of their combination of large population size and economic power. Additionally, because of their size and waves of immigration from other countries, these nations might have higher genetic diversity than smaller countries. Finally, there is considerable variation between the national medal counts in the summer versus the winter games. Tropical/desert nations rarely, if ever, win medals in the Winter Olympics. That is a clear bias!

On the other hand, if one viewed Finland's Olympic medal count against its population size, then Finnish athletes do extremely well. The ratio of total winter medals per million individuals for Finland is 85.5 compared with only 8.5 for the United States! Similarly, the summer medal count per million individuals for Jamaica is 26.8 and 16.3 for Kenya compared with only 7.6 for the United States. So, by these metrics, maybe the Finns, Jamaicans, and

Kenyans are among the greatest athletes in the world? Assuming that they might be, race clearly explains nothing.

Are Finns the world's best athletes? In 1924, one might have concluded that the answer was yes. But then the Olympics changed, and the answer to the question, as silly as it is, changed to no. What changed? The genetics of Finns did not change. What changed was culture and the political economy of Olympic sports.

DO AFRICANS RUN FASTER AND JUMP HIGHER THAN EUROPEANS AND ASIANS?

No. Although this trope is common and seems true to many, there simply is no data to support its veracity. Variation exists, but that variation is not racial. Here is why.

When people ask this question, they usually think about sprint events such as the 100-meter races that are run on land by elite athletes. However, we would not be pondering this question if we were thinking about sprint events in water (tables 6.1a and 6.1b), even though the necessary muscles are much the same.

World records in land sprint events are dominated by individuals whose ancestry is primarily from western and central African descent. Note that we said ancestry, not race. Ancestry is more specific. African Americans, most of whom do indeed have western and central African ancestry, also have

TABLE 6.1A
Male (M) and Female (W) World Record Holders: Land Sprint Events (< 400 Meters)

Event	Time	Name	Country	Ancestry
100 m (M)	9.58	Usain Bolt	Jamaica	African/European
100 m (W)	10.49	Florence G. Joyner	USA	African/European
200 m (M)	19.19	Usain Bolt	Jamaica	African/European
200 m (W)	21.34	Florence G. Joyner	USA	African/European
400 m (M)	43.03	Wayde Van Niekerk	RSA	European
400 m (W)	47.60	Marita Koch	GDR	European
110 m hurdles (M)	12.80	Aries Merritt	USA	African/European
100 m hurdles (W)	12.20	Kendra Harrison	USA	African/European
400 m hurdles (M)	46.78	Kevin Young	USA	African/European
400 m hurdles (W)	52.16	Dalilah Muhammad	USA	African/European

TABLE 6.1B
Male (M) and Female (W) World Record Holders: Water Sprint Events (< 200 Meters)

Event*	Time	Name	Country	Ancestry
50 m FS (M)	20.91	César Cielo	Brazil	European
50 m FS (W)	23.67	Sarah Sjöström	Sweden	European
100 m FS (M)	46.91	César Cielo	Brazil	European
100 m FS (W)	51.71	Sarah Sjöström	Sweden	European
50 m BFly (M)	22.27	Andriy Govorov	Ukraine	European
50 m BFly (W)	24.43	Sarah Sjöström	Sweden	European
100 m BFly (M)	49.50	Caeleb Dressel	USA	European
100 m BFly (W)	55.48	Sarah Sjöström	Sweden	European
50 m BS (M)	24.00	Kliment Kolesnikov	Russia	European
50 m BS (W)	26.98	Liu Xiang	China	Han Chinese
100 m BS (M)	51.85	Ryan Murphy	USA	European
100 m BS (W)	57.57	Regan Smith	USA	European
50 m BrS (M)	25.95	Adam Peaty	UK	European
50 m BrS (W)	29.40	Lilly King	USA	European
100 m BrS (M)	56.88	Adam Peaty	UK	European
100 m BrS (W)	1:04.13	Lilly King	USA	European

*BFly = butterfly, BrS = breaststroke, BS = backstroke, FS = freestyle.

on average 16 percent of genes resulting from their European descendants.[11] Jamaicans have slightly less European ancestry, about 12.4 percent, but still a good amount.[12]

Conversely, world records in swimming sprint events are dominated by persons of European descent. It is important to understand that the biomechanics of sprinting on land and in the water are not very different. Performance is strongly correlated with muscle strength in both land and water sprinting events. Swimmers tend to be stronger in the arms and shoulders than runners but equivalent in leg strength.[13] This makes sense, as it is much harder to propel oneself through water than through air. Thus, if we were to make a claim about which continental group is fastest, these data would support the contention that Europeans are faster than Africans. It takes more athletic strength to be an elite swimmer than a runner, and the world record-holding persons of African descent also have considerable European ancestry. Of course, this is a biased test, because it looks only at an infinitesimally small number, the top fraction of the total population, and does not say anything about the average or variation around the average.

One can also ask why are there so few elite U.S. African American swimmers. The answer is simple: racism and classism. The history of slavery and racial segregation in the United States has made it difficult for African Americans to gain access to pools, and safe places to swim as well as lessons and coaches. This pattern remains. In 2014, USA Swimming reported that only 5.3, 2.9, and 1.0 percent of its membership was Asian, Hispanic, and African American, respectively. This lack of participation of racial minorities in competitive swimming also has a bearing on the rate of accidental death by drowning in the United States.[14]

Because swimming is an important skill to prevent drowning, Joe and his wife made sure that their children took lessons. Soon after, their eldest son's smooth stroke was noticed by his coach, previously with the Canadian Olympic Swim program, and at that time of the Sun West Swim Club in Glendale, Arizona. Eventually, both of Joe's children became competitive swimmers. Joe's older son excelled at middle distances of 200 to 400 meters, particularly medley events. Joe's younger son was a sprinter, excelling at freestyle and the butterfly. Joe's dream of watching the U.S. flag go up with his sons on the medal platform at the 2020 Summer Olympics faded when his older son chose to pursue mathematics and his younger son switched to concert piano.

African American swimmers, including Cullen Jones, Lia Neal, Simone Manuel, and Reece Whitley, are beginning to make their mark in competitive swimming. Jones was the first African American to win a gold medal, in the 4 × 100-meter relay with Michael Phelps, Jason Lezak, and Garrett Weber-Gale at the 2008 Summer Games in Beijing. He took the silver medal in the 50-meter freestyle at the London games in 2012. Simone Manuel tied for the gold medal in the 100-meter freestyle with Penny Oleksiak (Canada) at the 2016 Summer Olympics, also earning a silver medal in the 4 × 100-meter freestyle relay (with Abbey Weitzeil, Dana Vollmer, and Katie Ledecky). At the 2017 World Championships, Manuel anchored the 4 × 100-meter relay (with Kathleen Baker, Kelsi Worrell Dahlia, and Lilly King), setting a new world record at 3:51.55. She also anchored the mixed 4 × 100-meter freestyle (with Caeleb Dressel, Zach Apple, and Mallory Comerford), with a world record of 3:19:97 in 2019. The Australians and Americans were dead even when Manuel entered the pool.[15] She totaled seven medals in that meet (two individual gold, two team gold, and three individual silver).

Finally, it should be noted that the specific sports at the Olympics and World Championships have been dominated by only a few countries in given periods, and these have changed over time. For example, swimming has been dominated by the United States and Australia for the last fifty years, but Japan and China are beginning to make inroads. China has earned forty-three Olympic swimming event medals since 1984. Since the 1990s, China has also begun to be a force in diving.

Similarly, female gymnastics was once dominated by teams from Eastern Europe and China, but this trend has been overturned by the United States, with African American gymnasts like Gabby Douglas and Simone Biles making major contributions. In addition, U.S. men once topped international basketball competitions, but now teams from western and eastern Europe are legitimate competitors, as noted by the number of stars from those areas who have signed NBA contracts.

A final nail in the coffin of "racial" claims of athletic dominance is the performance of various populations by gender. For example, U.S. men's soccer (football) has been disappointing, whereas the American women's team has dominated world competition for the last two decades. (The socially defined race compositions of these teams is similar.) Similarly, U.S. women's volleyball has maintained consistently higher rankings over the last fifty years compared with men's volleyball.

Popular mythology tells us that "white men can't jump." This is supported primarily by observations of NBA and NCAA basketball (but not of volleyball). Persons of European descent are rarely seen in the slam dunk competitions. Performances in this event by former NBA players Dominique Wilkins, Spud Webb, Michael Jordan, and Larry Nance (all of primarily African descent) are legendary. However, we cannot conclude from this anecdotal evidence that persons of European and East Asian descent are genetically less capable of jumping than those whose ancestry is primarily African. (Remember that all the individuals listed earlier also have European ancestry.)

Table 6.2 lists the male and female world record holders in the three track and field jumping events: high jump, long jump, and triple jump. Performances are dominated by persons of European and not African descent. Once again, we cannot conclude that superior jumping ability is found only in persons of African descent.

TABLE 6.2
Male (M) and Female (W) World Record Holders in Jumping Events

Event	Distance (m)	Name	Country	Ancestry
High jump (M)	2.45	Javier Sotomayor	Cuba	African/European
High jump (W)	2.09	Stefka Kostadinova	Bulgaria	European
Long jump (M)	8.95	Mike Powell	USA	African/European
Long jump (W)	7.52	Galina Christyakova	USSR	European
Triple jump (M)	18.29	Jonathan Edwards	UK	European
Triple jump (W)	15.50	Inessa Kravets	UKR	European

WHY ARE THERE SO MANY AFRICAN AMERICAN BASKETBALL PLAYERS AND LATIN AMERICAN BASEBALL PLAYERS?

Basketball was invented by Canadian-born James Naismith in 1891. Due to the racially segregated character of American society, almost all of the early basketball clubs were organized around ethnic groups. This included teams such as the Smart Set (African Americans from New York City), the Buffalo Germans (German Americans, Buffalo, NY), the Original Celtics (Irish Americans from the Hell's Kitchen area of New York City), and the South Philadelphia Hebrew Association (Jewish Americans from Philadelphia). In the 1930s, there was also a highly successful barnstorming team composed of Chinese Americans.[16]

Powerful evidence against the genetic racial theory of basketball includes the fact that as the social and cultural conditions of American society changed, so did the ethnic composition of top basketball players. Basketball took root in the inner cities of America and in the farm fields of the Midwest during the 1950s. Unlike tennis, golf, and swimming, basketball required little specialized equipment and space. In the 1980s, the sport began to spread from the Midwest to inner cities around the world. As an international nutrition postdoctoral student, Alan was surprised to find basketball courts in the remote village of the Solis Valley, Mexico.

Further evidence against the genetic racial theory of basketball is that the biomechanics required to become a great basketball player are very similar to those required to be a great volleyball player. Support for this claim comes from the fact that these athletes routinely sustain similar types of injuries resulting from movement during these sports.[17] Joe grew up playing basketball in the tough working-class, racially segregated urban/suburban environments of north central New Jersey. He played in community center club leagues and faced many athletes who went on to NCAA Division 1 and professional basketball careers.

During his years as a graduate student and assistant professor, Joe continued to play in pick-up games against Division 1 and professional basketball players, including NBA players B. J. Armstrong (Chicago Bulls), Scott Brooks (Minnesota Timberwolves and current Washington Wizards coach), and Richard Jefferson (New Jersey Nets). His team even won some games.

However, in college, he did not play basketball but learned to play volleyball, first as an undergraduate at Oberlin College and then as a graduate student at the University of Michigan in the Midwest Intercollegiate Volleyball Association (MIVA). At that time, the vast majority of MIVA teams were clubs and therefore did not play on the same schedule as the school's varsity male sports. So, Joe got to play against Division I schools (including Miami University of Ohio, the University of Toledo, Michigan State, Purdue, Northwestern, and Notre Dame) and nationally ranked teams such as Ohio State and Penn State. The point of this personal note is that Joe has direct experience to claim that basketball and volleyball skills are virtually identical and that the latter were harder to acquire than the former. Yet, if one looks at the demographics of these sports, men's basketball is disproportionately African American, and men's volleyball is disproportionately European American. On the other hand, it is telling that women's basketball and volleyball are more equally balanced by race and ethnicity. Finally, at the international level, volleyball dominance shows no association with socially defined race. Again, this demonstrates that performance in elite sports cannot simply be reduced to the athletes' racial origin.

Baseball

Joe's autobiography could well be called "The Thick Gray Line." That line refers to the gray bricks that made up part of the rear wall of McKinley Elementary school in Westfield, New Jersey. Almost every day of the summer, the kids of its ethnically mixed student body—Irish American, Italian American, and African American—played slow-pitch baseball with a tennis ball on the field behind the school. A fly ball that landed over the thick gray line was a home run.

The fifth and sixth graders hit home runs on that field at about the same frequency of major leaguers in professional baseball, and yes, they kept stats! However, even in slow-pitch, as a third-grader, Joe struggled to get on base. The pitchers threw great stuff, particularly curves and sinkers that would cause Joe to regularly ground or fly out.

Joe and friends also played fast-pitch stickball on the other side of the school. They played with sponge balls, baseballs shaped with seams, and regular baseball bats. This was not stickball in the classic sense of using a broomstick. They played with teams of three kids per side: a pitcher and two fielders. You could strike out, ground out, or fly out. A strike zone was marked on the wall in chalk. Hits were calculated by distance (singles, doubles, triples, and homers). Joe struggled in this game as well, but his younger brother, Warren, was outstanding. He went on to play varsity baseball at Westfield High School and then switched to its tennis team.

Many American kids of Joe's generation could describe a similar experience. Baseball has been called the "American game." It had been played since at least 1857. The National League (originally known as the National League of Professional Baseball Clubs) was founded in 1876. The American League was founded in 1901. Baseball spread to the Caribbean during the Spanish-American War in 1898. World War II spread baseball to Japan and the Pacific.

American baseball was never officially racially segregated. However, in 1887, one minor baseball league voted against offering contracts to Negro players, sending an important signal that Negroes were not welcome in major league baseball. Latinos (here defined as persons from North or Central America, the Spanish-speaking Caribbean, or Mexican Americans with a shared cultural heritage of colonization by Spain and use of the Spanish language) also began to appear in baseball in the 1880s. These athletes walked a precarious line of being acceptable to "whites" while participating in the Negro leagues.

Baseball's official color line fell in 1947, when Jackie Robinson signed with the Brooklyn Dodgers. The Boston Red Sox considered signing Robinson, but Jackie had a disastrous tryout with that team due to its racial attitudes. Indeed, one prominent Boston sportswriter suggested that the real "curse" of the Red Sox was not from trading Babe Ruth to the Yankees but the inability of the franchise to deal with its own and the broader city's racism.[18]

By 1968, major league baseball was composed of 15 percent African Americans and 7 percent Afro-Latinos. Like many major league sports, it suffered from "stacking," the assignment of players to fielding positions by socially defined race. African Americans outperformed European Americans by objective statistics by position, but they were overrepresented as outfielders and at first base and underrepresented as pitcher and catcher. The latter positions, like the quarterback in football, are thought to need greater cognitive capacity.

The percentage of African Americans in the major leagues declined steadily from this period. In 2018, only 8.2 percent of major league rosters were made up of African Americans. Conversely, the percentage of Latinos has grown steadily. By 2018, 28.2 percent of major league players were Latino, and these players are overrepresented at second base and as shortstops. Yet, they are also less likely to be drafted than European Americans.[19]

Once again, we are left with the question of whether representation in baseball or the stacking of positions results from genetically determined differences in socially defined races. We have established why this is highly unlikely. The genetic composition of people who are described as "Latino" is another example of strong evidence against this claim. Due to the history of colonialism and slavery in Latin America, the genetic ancestry of individuals varies considerably and included Amerindian, African, and European elements.[20] For example, Colombians have on average 64, 29, and 7 percent European, Amerindian, and African ancestry, respectively. Puerto Ricans have 72, 16, and 12 percent European, Amerindian, and African ancestry, respectively. Cubans range from being virtually all European with some Amerindian ancestry to virtually all African with some Amerindian ancestry.[21] Baseball players from all of the places and ancestry combinations have excelled at the sport. Once again, this exposes a supposed racial theory of baseball performance as patently ridiculous. The arenas of baseball and basketball show the social influence on sports participation and excellence.

ARE EUROPEANS STRONGER THAN AFRICANS AND ASIANS?

As with previous questions, people who ask are usually thinking about the performance of elite athletes in sports associated with strength, such as offensive linemen in American football or wrestling and weightlifting competitors. If we examine the Olympic world records in weightlifting, a striking pattern emerges for male and female athletes. Records are kept for three categories: snatch, clean and jerk, and total weight. Persons of East Asian

TABLE 6.3

Male (M) and Female (W) World Record Holders for Weightlifting (Total Weight) in 2019

Name	Ratio*	Country	Ancestry
Li Fabin (M)	5.21	China	East Asian
Shi Zhiyong (M)	4.97	China	East Asian
Lü Xiaojun (M)	4.67	China	East Asian
Jiang Huihua (W)	4.33	China	East Asian
Kuo Hsing-Chun (W)	4.17	Chinese Taipei	East Asian
Liao Qiuyun (W)	4.13	China	East Asian

Note: All females listed with a world record were from China or North Korea. For males, notable performance by persons not from China or North Korea were Lasha Talkhadze, Georgia (4.40 ratio, Eurasian), and Sohrab Moradi (1.80 ratio, Eurasian).

*Represents the ratio of the weight category to the athlete's body weight.

Source: Wikipedia, "List of World Records in Weightlifting," accessed April 29, 2021, https://en.wikipedia.org/wiki /List_of_world_records_in_Olympic_weightlifting.

descent—specifically from China and North Korea—dominate this sport, along with a few Middle Easterners, North Africans, and Europeans.[22] When we calculate the strength of individuals by the relationship of the total weight category to their body weight, we find that persons of Chinese descent win hands down (see table 6.3). There are no sub-Saharan Africans holding any of the world records. So, to answer this question, East Asians' elite performance in strength fares better than Europeans', with no evidence of world-record-level achievements by sub-Saharan Africans (or any other tropically derived population) in these events.

But, once again, we cannot ascribe the differences in performance in weightlifting to biological race. Not all East Asian nations claim records in weightlifting: there are no record holders from Vietnam, Laos, or Cambodia. The biomechanical requirements for elite-level weightlifting are associated with a shorter limb-to-height ratio. This provides a mechanical advantage to lifting heavy loads.[23] Thus, populations that evolved in climates that produced body proportions with shorter limb-to-height ratios would have an advantage in this sport. We know that human (and other large-bodied mammal) body proportions were influenced by adaptation to climate. Bergmann's rule states that for homeotherms (mammals, birds), body mass is inversely related to ambient temperature (as temperature goes down, body mass goes up).

In addition, Allen's rule states that species or populations that have adapted to warmer climates tend to have longer body extremities (feet and arms) compared with those who adapted to colder climates.[24] Thus, the combination of these rules strongly suggests that individuals with ancestry from warmer, tropical climates will not have body proportions that are conducive to elite weightlifting.

This is similar to what we learned about the advantage of persons adapted to high altitudes for excelling in sports that require endurance. Thus, the absence of athletes from populations of tropical ancestry from the highest levels of weightlifting competition is unsurprising. It should also be noted that other temperate-zone populations that are closely related to Chinese and North Koreans, such as Japanese and South Koreans, did not produce any world record holders in Olympic weightlifting either. This again indicates that there is a social and cultural aspect to achievement in this sport.

It is well established that physical strength is a heritable trait.[25] One recent study attempted to use GWAS to identify genetic variants associated with the greater strength of elite Russian weightlifters. The researchers found three SNPs that were statistically significant: rs120554409, G allele near the *MLN* gene that encodes promotilin, associated with contraction of smooth muscle; rs4626333, G allele near the *ZNF608* gene, which encodes a zinc finger transcription factor protein; and rs2273555, A allele in the *GBF1* gene, which encodes the Golgi-specific brefeldin A-resistance guanine nucleotide exchange factor 1 protein. They found in a follow-up study of sub-elite Polish weightlifters that GG homozygotes at rs4626333 SNP showed better competition results and greater cross-sectional type II muscle fiber area than the other genotypes.[26] The F_{ST} values for these SNPs worldwide are 0.103, 0.198, and 0.116, respectively, meaning that if variation alone in these genes were responsible for elite strength performance, then virtually every population would have lots of people who are capable of performing at elite levels of strength athleticism.

In summary, elite-level performance in strength-related athleticism is related to body proportions and muscle mass composition. Temperate-zone populations are more likely to have the combination of body mass and limb proportions for elite performance compared with those from tropical zones. The outcome of world competition in strength-related events is also influenced by social and cultural factors. Climate and culture are important. Race is not.

ARE ASIANS MORE FLEXIBLE THAN AFRICANS AND EUROPEANS?

This is a particularly hard question to answer, because there doesn't seem to be a generally accepted definition in the sport or medical literature for the physical trait known as flexibility. There seems to be some agreement that genetic variants might have something to do with the mechanical properties of musculoskeletal soft tissue that are associated with the range of motion of particular joints.[27] Alan's wife and daughter are hypermobile because of a genetic condition called Ehlers Danlos syndrome, which affects collagen. But that is a rare condition. The question of the greater flexibility of East Asians is often associated with claims of their superiority in sports such as gymnastics and diving, which require high degrees of contortion.

Once again, if we turn to the records of elite performance in the sports of gymnastics and diving, we find no data to support the conclusion of greater East Asian flexibility, as least as related to performance in gymnastics. Tables 6.4a and 6.4b present data for total medals in Olympic individual all-around competition. There's no evidence of dominance by East Asians in gymnastics. Indeed, Europeans of both sexes dominate these events. Japan's success in male gymnastics (thirteen medals) is countered by the failure of its women (no medals) and China's performance (only three bronze medals). The recent emergence of African American women Gabby Douglas, the 2012 Olympic all-around gold medalist, and Simone Biles, the 2016 gold medalist in vault, floor, and all-around as two of the finest female athletes in the history of this sport severely erodes this stereotype.

TABLE 6.4A
Total 2019 Olympic Medals for Women in the Gymnastics Individual All-Around Event

Country	Medals	Ancestry
USSR/Russia	20	European
Romania	8	European
Czechoslovakia	2	European
Germany	2	European
Hungary	2	European
USA	8	European (6), African American (2)
China	3	East Asian

Note: It is notable that the three medals won by China were all bronze.

TABLE 6.4B
Total 2019 Olympic Medals for Men in the Gymnastics Individual All-Around Event

Country	Medals	Ancestry
USSR/Russia	19	European
Japan	13	East Asian
France	6	European
Switzerland	6	European
Italy	5	European
Germany	4	European
Finland	3	European
USA	2	European

Note: A number of countries earned only one medal, including China, South Korea, Hungary, Yugoslavia, and Austria.

Diving

The data for Olympic medals in diving (10-meter platform, synchronized 10-meter platform, 3-meter springboard, and synchronized 3-meter springboard) mirror those found for gymnastics (tables 6.5a and 6.5b). Europeans have long dominated the sport in total medal counts. Chinese divers have recently started to earn medals, particularly in synchronized diving, but other East Asian nations—notably Japan and South Korea—have not done well.

In summary, all that we know about the continuous nature of human genetic variation and how it affects physical traits does not support the idea that there are genetically racial differences in flexibility. East Asians show no dominance in gymnastics. In addition, both diving and gymnastics are associated with specific cultural and social norms throughout the world.

TABLE 6.5A
Total Medals for Women in 2019 Olympic Diving Events (Top Five)

Country	Medals	Ancestry
USA	44	European
China	35	East Asian
Russia/USSR	15	European
Germany	15	European
Sweden	8	European

Note: Notable countries not making the top five were Canada (7), Australia (6), Malaysia (2), and Mexico (3).

TABLE 6.5B
Total Medals for Men in 2019 Olympic Diving Events (Top Five)

Country	Medals	Ancestry
USA	73	European
China	34	East Asian
Russia/USSR	17	European
Germany	17	European
Mexico	11	European, Amerindian, African

Note: Notable countries not making the top five were Italy (9), Sweden (6), and Australia (5).

In the United States, African Americans have been excluded from both sports until recently. The emergence of a new generation of African Americans in gymnastics (but not yet in diving) at the elite level is evidence against racial theories.

DO AFRICANS HAVE THICKER BONES, EXTRA MUSCLES, OR EXTRA TENDONS?

No.

All human populations are anatomically the same with regard to the number of muscles and tendons. Bone density varies by population but not socially defined race. It is a complex trait, influenced by genetics, epigenetics, endocrine function, and environmental factors (especially nutrition and disease). A variety of hormones positively influence bone density including insulin, growth hormone, insulin-like growth factor I, estrogen, testosterone, vitamin D_3, and calcitonin. Parathyroid hormone, cortisol, and thyroid hormones will negatively impact bone growth.

Much of what we know about variation in bone density has been driven by studies of bone-associated diseases, such as osteoporosis or risk of fracture injuries. These studies were organized within the socially defined race paradigm that is still utilized in medicine. For example, a 1995 study of bone mineral density (BMD) in Black and white men took measurements in the lumbar spine, femoral neck, and radial shaft.[28] The study, which used only 34 Black and 160 white men, controlled for education level, smoking, exercise, and fractures. The researchers found no significant Black/white differences in mean weight, height, or body mass index (BMI). They also found that the Black men had higher BMDs than the white men at every site: 5 percent greater for the radius, 10 percent greater for the lumbar spine, and 20 percent greater for the femoral neck. The study concluded that Black men had higher BMD than white men and that this was not due to greater body size. The investigators also concluded that the lower hip fracture risk reported for Black than for white men was not due to a difference in hip axis length; rather, it was the result of greater BMD in the former group.

Studies such as this contributed to the false belief that Blacks have always had greater bone density than whites. A more recent study demonstrated that this was incorrect.[29] The motivation for the study was the fact that reference populations from the United States are often used around the world for representative measures of BMD by sex, age, and race. Mukwasi and colleagues examined BMD in adult Black Zimbabwean women and compared it with that of Black and white U.S. women. They examined 289 participants age 20–69 years, controlling for BMI. In the 20–59-year age range, compared with U.S. white women, mean BMD for Black Zimbabwean women was 4.5 to 7.4 percent lower for the lumbar spine but 2.0 to 4.8 percent higher for the total hip and 0.2 to 10.2 percent higher for the femur neck. The authors also compared the Zimbabwean with U.S. Black women of ages 20–59 and found that the mean BMD for the former was 9.1 to 11.5 percent lower for the lumbar spine and 1.4 to 8.1 percent lower for the total hip. Black Zimbabwean women also had lower mean weight and BMI compared with U.S. women. Thus, from these results we learn that there is no simple way to generalize or predict the bone density of human populations.

In addition, given the complexity of the determinants of bone density, the genomic foundation of this trait has been shown to be just as complex. Recent GWAS identified more than one hundred genetic loci associated with bone density.[30] In addition, given that bone density is profoundly influenced by the conditions under which an individual develops, epigenetic influences have been demonstrated.[31] Thus, given the enormity of genetic variation on the continent of Africa, along with the tremendous variation in environmental circumstances that influence bone development, there is no simple way to claim or predict a specific bone density for any African population.

CONCLUSIONS

Since the 1940s, people have seemed to adhere to the stereotype of the "dumb jock." Athletic ability and intelligence seem to be polar opposites among individuals. We didn't fit that stereotype. Alan was a halfback, and Joe was a wide receiver in junior high school. Alan's high school team was undefeated and co-state champions. In college, Joe went on to excel in a variety of sports: basketball, rugby, softball, and volleyball. He was also his college's chess champion and still is a highly rated United States Chess Federation player. Alan was a member of his college's intramural football team, which won three consecutive championships. Having never played lacrosse until college, he made the varsity lacrosse team (ranked eleventh in the nation). Alan is five foot five and was likely the shortest and lightest-weight person on all of these teams.

As harmful as the stereotype of the dumb jock is, it is made worse when racialized.[32]

Chapter Seven

INTELLIGENCE, BRAINS, AND BEHAVIORS

In this chapter, we dive into what is possibly the most destructive myth of race: that there are innate differences in intelligence among racialized groups. This is often considered the flip-side of believing that races differ in athletic ability. We examine what is meant by intelligence, how it is measured, and what we actually know about the genomics of intelligence and cognitive abilities. We also address related questions about behaviors and traits including personality, mental health, violence, and criminality. The bottom line is this: there is absolutely no evidence that differences in any of these traits or behaviors have to do with genetic differences among social races. Any thoughts that there are racial differences in intelligence due to genetics are hurtful and false.

WHAT IS INTELLIGENCE?

Not surprisingly, there is no general agreement on how to define the complex trait that we often reduce to one word: intelligence. Is intelligence a thing that you have or don't have, or have in varying degrees? Or are there multiple types of intelligence that might or might not be related to each other?

Three definitions of and frameworks for thinking about intelligence are commonly used. The first is *Cattell-Horn-Carrol* (CHC) theory. This holds that there are three strata of intelligence, which are hierarchically linked: narrow abilities (I), broad abilities (II), and general ability (III). Stratum III is referred to as general intelligence or *g*. The second stratum, fluid ability, allows one to cope with novelty and to think rapidly and flexibly. Finally, crystallized or narrow ability refers to the general store of knowledge.

The second commonly used concept is *multiple intelligence theory*. This holds that intelligence resides in eight or so different components, including linguistic, mathematical, spatial, musical, bodily/kinesthetic, naturalist, interpersonal, and intrapersonal.

The third commonly used definition is *triarchic theory*, which holds that intelligence consists of three components: creative, analytical, and practical.[1]

There are variations on all of these themes. The key point is that even though we often think of intelligence as a single attribute, everyone knows someone who is gifted at some tasks (say, being good at directions, memorizing lines, or hitting a golf ball) and at the same time very bad at others. All themes highlight various forms of intelligence.

DO RACES DIFFER IN THEIR COGNITIVE ABILITIES AND INTELLIGENCE?

As stated in our responses to previous questions in this book, humans do not have biological races, and therefore there can be no genetically based racial differences in intelligence. However, some have reframed the question, avoiding the racial claim and changing the focus to genetically defined populations. Again, we answer "no" to that reformulation. Intelligence, however defined and measured, is the wondrous result of a complex mix of multiple genes, chance events during development, and environments.

Most claims of racial differences in intelligence refer to the CHC definition; specifically, generalized intelligence or *g*. Intelligence testing has a long history in the United States, beginning with conversion of the intelligence quotient (IQ) score into a tool for eugenics. In 1917, Henry Goddard published a paper titled "Mental Tests and the Immigrant."[2] He administered IQ tests to steerage passengers at Ellis Island and determined that those entering could be classified as "normal," "borderline," "feeble-minded," "moron," or "imbecile." He found that these categories differed by nationality, with just 10, 0, 7, and 0 percent of Jews, Hungarians, Italians, and Russians classified as "normal," respectively. Conversely, over 83, 80, 79, and 87 percent of immigrants from those countries were thought to fall into the categories of "feeble-minded" or worse. Seven years later, eugenicist Lothrop Stoddard reported to the U.S. Congress that more than 46 percent of American

Negroes from the North and 75 percent of those from the South displayed inferior intelligence.[3]

Discussion of the intellectual inferiority of European immigrants shifted through the early portion of the twentieth century, particularly as these groups became "white." However, the IQ test scores for African Americans improved more slowly. As early as 1932, IQ testing had shown a fifteen-point (about 1 standard deviation) difference between whites and Blacks in the United States.[4] In the 1990s, Richard Herrnstein and Charles R. Murray argued in *The Bell Curve* (1994) that this IQ test gap had been stable throughout the twentieth century, also observing that East Asians scored about three points higher on average than Europeans and that Latinos fell between Europeans and Africans.[5] Throughout this narrative of IQ test difference was a constant claim that the IQ test score was a singular, unbiased measure of intelligence (g) and that the racial IQ test gaps were driven primarily by genetic differences influencing cognitive performance. However, the history of IQ testing is far from unbiased, and we have already discussed the numerous fallacies of believing that there are genetically based racial groups in our species (chapter 3).[6]

IS THERE A GENETIC BASIS FOR INTELLIGENCE?

Yes, but. . . . it is wrong to say that intelligence is caused by genes. How genetics is supposedly causally linked to degrees and types of intelligence is complex and varies for each person. And, of course, this has nothing to do with socially defined race.

Intelligence, however defined or measured, is related to brain function. Specifically, it is related to the function of the prefrontal cortex, neocortex, and superior parietal, temporal, and occipital cortexes. In addition, the subcortical regions of the brain, such as the striatum, and the integration of the parietal and frontal lobes are especially important. The complexity of gene involvement is indicated by the fact that more than 82 percent of the estimated total of 23,000 human genes are expressed in the brain, nearly double the 46 percent of genes that are expressed in other tissue types. However, even this is an underestimate of genetic contributions to cognitive function, as a wide variety of physiological systems outside of the brain also contribute to what we call intelligence. In addition to genetic influence on cognitive function, a large number of environmental factors influence brain function, such as nutritional state, exposure to neurotoxins such as lead, and infectious disease.

The heritability of intelligence has been consistently shown to be between 0.50 and 0.80; in other words, greater than 50 percent of intelligence is generally attributable to heredity. Estimates of heritability also differ for components: verbal aptitude (0.70) seems to be higher than spelling (0.50) and mathematical aptitude (0.30).[7]

However, these heritability estimates should be taken with a large lump of salt. Heritability, or h^2, is typically measured in humans by comparing similarities between monozygotic and dizygotic twins. As noted earlier, monozygotic twins share all genomes, whereas dizygotic twins share half of their genomes. Studies of twins tend to overestimate the genetic contribution attributable to the unintended correlation of twin environments. First, identical twins share the same physical attributes, so any aspects of the environment associated with how one is treated due to physical appearance are shared. Also, of specific importance for identical twins socially defined as white in America, there are rigorous requirements concerning the attributes of potential adoptive parents. Thus, such twins tend to be adopted into families with shared socioeconomic characteristics. So, we can say with confidence that there definitely is a genetic component to cognitive performance, but cognitive performance as a complex trait is also strongly influenced by environmental conditions. Finally, it is really important to understand that stating that intelligence is heritable is not the same as saying that the genetic basis of the trait must differ among socially defined races, as many people who should know better do, including Murray.[8]

At present, GWAS of intelligence (measured by various IQ tests) predicts only 1–4 percent of the variation in the trait.[9] Such genome-wide association studies also have failed to discover many genomic variants that are consistently associated with the trait even within populations of the same ancestry. For example, a study examining 549,692 SNPs from 3,511 unrelated adults in the Lothian Birth Cohorts (1921, 1936), Aberdeen Birth Cohort (1936), and the Manchester and Newcastle Longitudinal Studies of Cognitive Aging Cohorts showed that none of the individual SNPs they discovered showed a replicable genome-wide association. One gene with a suggestive association with fluid intelligence was statistically significant in a binding protein (FNBP1L). However, this result was not replicated in the Norwegian Cognitive Neurogenetics sample.[10]

A 2015 GWAS examined exonic (genes that are expressed) variation associated with extremely high intelligence (*HiIQ*). *HiIQ* individuals were selected from the Duke University Talent Identification Program (TIP). This study used the top 1 percent of this group who had very high IQ scores. A control group was generated from the Minnesota Twin Family Study: 3,253 individuals identified as white with IQ scores distributed between 70 and 150. This resulted in just a single nonsynonymous SNP in the PLXNB2 locus, whose gene product has been associated with neuronal migration, explaining only 0.16 percent of the variance in IQ between the controls and the high-IQ group.[11]

Attempts to improve the power of GWAS to discover genomic variants associated with cognitive function have utilized proxies for the intelligence (g), such as educational attainment, or highest-level math course taken. There are obvious problems with these traits as proxies for cognitive function, as they clearly are influenced by differing opportunities and circumstances. However, even if we accept these limits, these new indicators perform only slightly better than IQ tests. For example, a study of a European population demonstrated that 74 genetic markers (SNPs) contributed to individual differences in educational attainment and explained about 20 percent of the variance in education levels. However, even with a 20 percent genetic contribution to educational attainment, if the effects are additive (and equal), we would expect that each SNP contributed only 0.27 percent to the variance in the trait.

In another large GWAS of Icelandic people, it was shown that 35 of 120 significant SNPs had a strong association with age of first child for women. The SNP rs192818565 had the highest association with lower educational attainment but was highly associated in women with having more children and having them at an earlier age. This suggests that what is at play here is not intelligence; rather, it is that female educational attainment is often negatively associated with having more children.[12]

One of the largest studies to date showed that 1,271 SNPs accounted for 11 percent of the variance in educational attainment (EA) in a sample of 1.1 million individuals of European descent. These results suggest that the roughly 1,200 SNPs explain as much variation in EA as family socioeconomic status (SES). They showed that 225 SNPs were significantly associated with cognitive test performance, 618 with self-reported math ability, and 365 with hardest math class taken.[13] Again, if we accept these results as true, they really say only that genes and environmental factors are equally important in determining someone's educational attainment. This is not a shocking conclusion.

Finally, one of the few genome-wide association studies that examined a cognitive trait in populations of different ancestry examined cognitive flexibility in a sample of more than 3,000 African Americans and 2,500 European Americans.[14] The variants found to be associated with this trait in African Americans were not significant in European Americans (and no significant SNPs were found for this group.) Nongenetic factors in this study were actually more significant than the SNPs. Older age and recent tobacco use were shown to make performance worse (for both groups); and years of education made performance better for both groups. Frankly, given the small size of the cohorts examined, it is amazing that this study found any significant genetic variants. Of course, the results were not replicated in these socially defined groups or in any other population. Therefore, after this long trip through many studies, we cannot make much of the results. Effects of single genes are small and typically fail to repeat across studies. Given what we know about genes and brains, that's actually not surprising.

Taken as a whole, these GWAS results on cognitive function support that idea that although genes are associated with individual IQ, the relation is extremely complex. Furthermore, the relationship between intelligence and genetics is likely to vary by population (due to different patterns of linkage disequilibrium). And even though there is a real but variable genetic contribution, the environmental contribution is just as real and equally, if not far more, important. Finally, whatever we now know about the intelligence and genetics is independent of and unrelated to social race. More on this in the next question.

DO GENES ASSOCIATED WITH INTELLIGENCE DIFFER AMONG RACES?

No.

A mainstay of psychometrics is the assertion that intelligence differs among biological races. Past reasoning was that because human groups differed in some physical traits, or at least were thought to differ, they must also differ in their intellectual capacity. When it became clear that the physical traits used to ascribe difference (such as skull angle and volume) did not differ in a way consistent with notions of racial classification, new tools to measure intelligence were invented. However, once studies of genetic markers in humans demonstrated the fallacy of grouping human beings into biological races, many psychometricians continued to ignore the genetic data, suggesting genetic differences accounting for intelligence among Blacks, whites, and Asians.

Recently, Harvard geneticist David Reich alluded to the possibility that genetic variants exist that are differentiated by population and that might account for differences in intelligence. In making this claim, he directly dismissed Joe's analysis suggesting that the measured IQ differences in human populations cannot result from genetic differences among them.[15] His dismissal was rooted in the idea that natural selection can act on complex traits (such as intelligence).

On this point we agree. Yes, natural selection can act effectively on complex traits; indeed, Joe spent his early career demonstrating it with regard to life history traits in fruit flies. What we question is whether this ever happened in the human species for the quantitative trait we call intelligence. There are two ways that natural selection could have accounted for altering the genomic foundation of cognitive performance in humans: there could have been direct selection for improved performance, or improvement could have resulted from selection on some other trait (correlated selection).

It is generally agreed that the greater intelligence of humans compared with other animals is the key adaptation of our species.[16] Our evolutionary lineage was characterized by increased brain size and complexity. Our skulls changed in ways that decreased the size of the face, specifically bones and muscles that allow for powerful chewing, and allowed more space for larger brains. Furthermore, it is argued that the primary driving force of all primate intelligence (including our own) is social interactivity.[17] Thus, most of the evolution of our intelligence capacity occurred long before anyone left Africa. This raises the question about whether direct selection for greater intelligence occurred after some humans left Africa, particularly for life in Eurasia.

Two theories have been used to claim greater natural selection for intelligence in Eurasia: the so-called winter hypothesis and r versus K selection.[18] According to the winter selection hypothesis, which dates to the turn of the twentieth century, the unique problems of winter (such as shelter, food storage, and lack of food) required greater cognitive sophistication to solve than life in the "easygoing" tropics, resulting in natural selection for greater cognitive capacity in Eurasians. The problem with this theory is that sub-Saharan Africans did not stay put in one place while Eurasians roamed the world solving new problems. Africa has eight climatic zones (equatorial, humid tropical, tropical, Sahelian, desert, and Mediterranean). The distribution of these zones differed in the Pleistocene period, but migration and survival in any of the zones required solving new problems as well.

The r/K selection theory, a variant of the winter selection hypothesis, was championed by Canadian psychologist J. Phillipe Rushton (1943–2012). It posited that humans could be arrayed along a spectrum of reproductive investment, which in turn was in opposition to somatic investment. Africans were r-selected (high reproduction, low investment in brain mass, lower altruism, lower social organization), and Eurasians were K-selected (lower reproduction, higher investment in brain mass, higher altruism, higher social organization). Rushton utilized this model to explain why Blacks were not found in high-IQ professions as well as various social pathologies that he claimed existed in Black societies worldwide. The problem with Rushton's theory was that he applied it backwards. The theory actually predicted that species in the temperate zones should be r-selected and those in the trop-ics should be K-selected. Moreover, it is a misapplication of an ecological idea applied across species to variation within a species. Finally, that eco-logical idea has been challenged, because it does not work in a wide array of situations.

Therefore, no one has presented a credible model for why some people should have evolved greater cognitive performance than others. Without a direct theory of selection, one is left to explain that evolution of greater intel-ligence in Eurasians resulted from an accident of evolutionary history. This approach does not work either. Methods have been developed to assess the signature of natural selection on complex traits.[19] Classic quantitative genet-ics tells us that selective sweeps reduce the heterozygosity of nearby neutral polymorphisms and provides a strong genetic signature of selection.[20] How-ever, in organisms like us, there is increasing evidence that "soft" selective sweeps, particularly adaptation resulting from changes in allele frequen-cies of existing genetic variation, could be the major mechanism of adap-tive events.[21] It is particularly hard to detect these signals in humans due to incomplete knowledge of our genetic architecture and weak selection signals of individual loci.

To overcome these difficulties, investigators in one study utilized a combi-nation of F_{ST} and branch analysis on thirty-eight complex traits to determine if there was evidence of directional selection in humans.[22] *Branch analysis* refers to examination of the lengths and patterns of branches in phyloge-netic trees of the genes that are undergoing positive natural selection among the populations being studied. The traits fell into six categories: quantitative physical traits, quantitative physiological traits, inflammatory or autoim-mune disorders, mental disorders, cancers, and miscellaneous traits related to menstruation. This study concluded that (a) natural selection on standing variants associated with complex traits is common in humans; (b) signatures of selection for any particular trait that is enriched in different human lineages indicating recent adaptation are most likely associated with temporary and geographically specific environmental challenges (e.g., malaria adaptation); and (c) there is a strong correlation between the strength of selection and how much the trait has changed (such as height and some physiological traits). The study also found that for traits without correlation between the estimated strength of selection and how much the trait changed (thirty-four of the thirty-eight traits they studied), selection on the trait was acting through selection on a correlated trait.

The second and third conclusions have particular relevance in considering whether consistent directional selection for any complex trait is possible. These data indicated that the genetic response to natural selection (change in allele frequencies) could be highly driven by chance, meaning that we should not expect any consistent or predictable changes for allele frequencies over the course of human evolution. This is also due to the fact that the evolutionary forces impacting any specific trait are heterogeneous (multiple mechanisms operating at the same time). These conclusions are entirely consistent with our assertion that it is highly unlikely that directional selection for greater or lesser intelligence, no matter how defined, drove observed differences in this trait among supposed human races.

Therefore, if no genes have been identified to date that differentiate human populations (or socially defined races) for cognitive function, why are there observed differences in IQ? The answer to this question is the same as the one we provided for health disparity. Cognitive performance is a complex trait with a strong environmental component. The environments of America's socially defined races have never been equalized. This has particular significance (as we have explained in a number of scholarly works) for estimating the genetic effects on any complex trait.[23] Any estimate of a genetic effect on a physical trait will be incorrect if the groups being compared are not reared in the same environment. This is an elementary principle of quantitative genetics, one that psychometricians have never bothered to heed when making claims about the genetic basis of intelligence differentials in socially defined races.[24]

In summary, the persistent belief that there are racial differences in intelligence are based on a series of fallacies and unsupported assumptions, including these:

(a) races are genetically real,

(b) IQ (*g*) measures intelligence (assuming that we truly understand the meaning of this term), and

(c) IQ (g) differences among socially defined races are measured in an unbiased and consistent way.

All of these assumptions are highly problematic and unsupported. Instead, it is clear that cognitive ability, like most traits that result from contributions from genes, environments, and chance, varies among the individual members in any group. However, intelligence is a requirement of all cultures. Cultures are all complex, and success within them often requires the effective use of one's intellectual ability.

DOES MUSICAL ABILITY DIFFER BY RACE?

Musical ability is considered one of the components in the multiple theory of intelligence. Similar to other complex traits, musical ability is hard to define. However, some biologic traits are associated with musical ability that display genomic variation within our species. What we know so far coincides with other complex traits: variation within populations exists, but we have no evidence that there is significant variation for these traits among populations or socially defined races.

Evidence for genetic contributions to musical ability can be found in unusual conditions, such as tone-deafness, the inability to detect notes that are out of key; and the opposite condition, absolute pitch (AP), the ability to identify and reproduce a musical note without reference to an external standard. Both conditions are rare in human populations (tone-deafness occurs in about 3 percent), and their genomic foundation is therefore very different from the normal range of musically related traits in humans.[25] Here we note that genes might be the foremost reason for variation in great ability or inability (the extreme ends of the spectrum) but not necessarily the small variation observed in almost everyone else.

Evidence for the genetic foundation of tone-deafness comes from the fact that it is clustered in families. A candidate gene analysis examining Europeans and East Asians found that it was associated in the former with variation at chromosome position 8q24. Studies of families in Finland using tools such as the Karma music test, Seashore pitch test, and the Seashore rhythm subtest found heritability performance on these tests at 0.42, 0.51, and 0.21, respectively. Other studies in this population found genetic associations for musical abilities with chromosome positions and specific genes.[26] Another study of musical ability in Finnish musicians found evidence of positive selection in another set of genes.[27] Yet a further study of the ability to discriminate pitch, duration, and sound patterns in seventy-six Finnish families found significant genetic associations at position 3q21.3 (SNP rs9854612), which is upstream of the *GATA2* (GATA binding protein 2) gene. This gene is involved in regulating the development of cochlear hair cells and the interior colliculus. The strongest genetic association was for sound patterns at position 4p14 (rs13146789, rs13109270) within the protocadherin (*PCDH7*) gene, which also regulates cochlear and amygdaloid complexes.[28]

The combination of the heritability of performance on these musical tests and evidence that so many of the candidate genes are associated with cognitive processes suggests that for these families, there is an actual genetic component to the variation in their musical abilities. However, it is critically important to understand that these estimates come from a set of families in Finland, a country whose genetic composition is strongly influenced by genetic drift (see chapter 6). Therefore, there is absolutely no reason to believe that these genetic associations would replicate across other Europeans, let alone the rest of the world.

For example, the frequency for the SNP rs13146789 associated with the *PCDH7* gene varies between 0.473 in Siberians and 0.213 in Koreans, and rs13109270 varies from 0.475 in northern Swedes to 0.228 in Koreans in dbSNP.[29] These values indicate that there should be adequate variation in all of these groups to allow for people with the musically associated phenotypes that might result from these variants. In summary, although genetic variation associated with musical ability has been discovered through modern genomic techniques, there is no reason to believe that there are significant differences in the genomic foundation of this trait among populations or socially defined races.

DO RACES DIFFER IN PERSONALITY?

Now do you think that's the wrong question to be asking?

We don't want to sound like a broken record, but personality is extremely difficult to define and therefore extremely difficult to measure. Definitions of personality commonly highlight variations in enduring patterns and thoughts, feelings, and behaviors that tend to allow an individual to respond in individually specific ways to specific circumstances. An individual's personality is associated with their cognitive processes, just as intelligence is a complex phenotype influenced by genetic, epigenetic, and environmental factors. In the case of personality, it is also important to focus on the components of environment, such as its physical and social/cultural aspects. Much of the biomedical literature focuses on aspects of personality that are associated with mental illness, whereas the literature in industrial psychology tends to focus on personality with regard to its application to improving institutional effectiveness.[30] In either case, stereotypical views of personality associated with socially defined race are disastrous.

The five-factor personality model is widely used in both biomedical and industrial psychology. This model includes scales associated with neuroticism or emotional stability, extroversion, openness to new experience, agreeableness, and conscientiousness. The scales also have subdimensions. Subdimensions of neuroticism/emotional stability include self-esteem, low anxiety, and even temperedness; extroversion includes dominance and sociability; openness, agreeableness, and conscientiousness subdimensions are achievement, dependability, degree of caution, and desire for order.

Table 7.1 summarizes data from a meta-analysis of the five-factor personality scale conducted with a large sample of individuals (> 300,000) from five socially defined races in America.[31] On positive personality metrics, whites and East Asians tended to score highest, and Blacks and American Indians tended to score lowest. Despite the large sample, there are serious issues with the design of this study. Specifically, the data used in this study were distributed among individuals from the general population (50 percent), business occupations (31 percent) and students in colleges (19 percent). In addition, the distribution of individuals in these categories by socially defined race is unequal. The data also indicate that there is a great deal of representation by individuals from each of the socially defined races in each score category in each component of the five-factor system. For example, for the dependability trait, there is virtually no difference among socially defined races. However, for traits having to do with need and desire, East Asians scored much higher than whites, who scored the same as Hispanics and Blacks.

An aspect of personality that has important implications for how socially defined race is lived in societies is *social dominance orientation* (SDO).[32] SDO is related to the order and dominance subscales of the five-factor personality scale, but it is designed to evaluate an individual's comfort with

TABLE 7.1
Five-Factor Personality Scale Ratings for Americans by Socially Defined Race

Factor	W-B	W-As	W-H	W-AI	Order
Emotional stability	−0.90	−0.12	0.03	−0.21	H > W >A > AI > B
Global measures	−0.12	−0.16	−0.04	ND	W > H > B > A
Self-esteem	0.17	0.30	0.25	ND	A > H > B > W
Low anxiety	−0.23	0.27	0.25	ND	A > H > W > B
Even tempered	0.06	−0.38	0.09	ND	H > B > W > A
Extroversion	−0.16	−0.14	−0.02	−0.21	W > H > A > B > AI
Global measures	−0.21	−0.07	0.12	ND	H > A > W > B
Dominance	−0.03	−0.19	−0.04	ND	W > B > H > A
Sociability	−0.39	−0.09	−0.16	ND	W > A > H > B
Openness	−0.10	0.11	−0.02	ND	A > W > H > B
Agreeableness	−0.03	0.63	−0.05	−0.28	A > W > B > H > AI
Conscientiousness	−0.07	0.11	0.08	ND	A > H > W > B
Global measures	0.17	0.04	0.20	ND	H > B > A > W
Achievement	−0.03	0.14	0.10	ND	A > H > W > B
Dependability	−0.05	−0.01	0.00	ND	W = H > A > B
Caution	0.16	ND	ND	0.25	AI > B > W
Order	0.01	0.50	0.00	ND	A > B > H = W

Note: The values listed are proportions of the standard deviation in the distribution of the response. All groups are compared with whites, allowing the ranking of nonwhite groups relative to one another. Whites, East Asians, Hispanics, American Indians, and Blacks rank first or second at eight, nine, ten, one, and five times, respectively. Whites, East Asians, Hispanics, American Indians, and Blacks rank fourth or fifth at four, three, two, three, and nine times, respectively. For American Indians, low ranking on these personality variables occur in three of the four studies in they were included.

AI = American Indian, A = East Asian, B = Black, H = Hispanic, ND = no data, W = white.

Source: H. J. Foldes, E. E. Dueher, and D. S. Ones, "Group Differences in Personality: Meta-Analyses Comparing Five US Racial Groups," *Personnel Psychology* 61 (2008): 579–616, https://onlinelibrary.wiley.com/doi/abs/10.1111/j.1744-6570.2008.00123.x.

social dominance. SDO survey instruments contain items questioning an individual's attitudes about gender and sexism, ethnic and racial prejudice, nationalism, cultural elitism, political and social conservatism, distribution of resources, meritocracy, military policy, punitive policies, social welfare, civil rights, and environmental policies. SDO has been validated across cultures and tends to show consistent results, with men displaying greater SDO than women and members of dominant racial and ethnic groups showing higher SDOs than subordinated populations.

For example, in Taiwan, SDO was significantly correlated with an individual's sexism, traditional attribution for poverty, attribution of economic inequality, support for the existing political system, support for the military and police, support for hegemonic nations, and opposition to college students. In Canada, SDO had significant correlations with sexism, ethnic prejudice, opposition to gay rights, environmental protection, punishment for police brutality, more equalitarian distribution of resources, support for capital punishment and more defense spending.[33]

In summary, we once again conclude that personality is a complex trait whose components are far from agreed on. Studies of personality routinely show differences and similarities in these metrics among socially defined races. Certainly, the overlap between the distribution of these metrics demonstrates that one cannot use socially defined race as an accurate predictor of the personality traits of any individual.

DO GENES ASSOCIATED WITH PERSONALITY DIFFER AMONG RACES?

Not surprisingly, as with intelligence, personality as measured by a variety of means has been shown to be heritable. The range for h^2 reported in the recent literature varies from 0.17 to 0.65. The higher estimates tend to come from more biomedically related mental illness metrics.[34] However, similar to the case with intelligence, the existence of genetic variation that influences various aspects of personality does not mean that a genetic basis for the distribution of such aspects exists among socially defined races. So far, genome-wide association studies that have attempted to find a genomic foundation for various components of the five-factor scale have been limited in scope and found few positive associations (table 7.2). Table 7.2 shows that eleven SNPs were identified as statistically significantly associated with five-factor traits from ten studies of Europeans and East Asians.

Even these minimal results need to be taken with yet another large grain of salt. One major difficulty that is immediately apparent with these data is that the social/cultural environments of these groups (as well as physical environments) are not equalized. Thus, even if we were to believe these results, we have no way of really determining the robustness of their effects across environments and cultures. Thus, similar to the GWAS results for intelligence, we know there is a genomic contribution to the personalities of

TABLE 7.2
GWAS Meta-Analysis Results for Five-Factor Scale Ratings for Various Populations

Trait	SNP	Gene	Population	Size	Replicated?
Agreeable	None		Sardinians	3,572	No
Agreeable	rs9650241, rs2701448, kgp6080068	Chromosome 8	Europeans	4,595	No
Agreeable	None		East Asians	3,898	No
Agreeable	rs8015351	RPS29	Koreans	4,919	No
Conscientious	None		Sardinians	3,572	No
Conscientious	rs2576037		Europeans	17,375	No
Conscientious	rs912765, NS.	LMO4	Koreans	4,919	No
Extroversion	None		Sardinians	3,572	No
Extroversion	rs57590327, rs2164273, rs6481128, rs1426371	GBE1, MTMR9, PCDH15, WSCD2	Europeans	122,886	No
Extroversion	rs12537271, NS	PTPN12	Koreans	4,919	No
Excitement seeking	rs7600563	CTNNA2	Europeans	7,860	No
Openness	None		Sardinians	3,572	No
Openness	rs1477268, rs20332794	RASA1	Europeans	17,375	No
Openness	rs16321695	IMPAD1	Koreans	4,919	No

Note: The limitations of the studies include the fact that only Europeans and East Asians were sampled (and certainly not all groups within these regions of the world) and that the studies were not replicated.

NS = not statistically significant.

The data are summarized from B. H. Kim, H. N. Kim, S. J. Roh, et al., "GWA Meta-Analysis of Personality in Korean Cohorts," *Journal of Human Genetics* 60, no. 8 (2015): 455–60, https://www.nature.com/articles/jhg201552.

individuals but no reason to suppose that the genomic foundation of personality differs among populations or socially defined races.

DO GENES ASSOCIATED WITH MENTAL ILLNESS DIFFER AMONG RACES?

At the risk of sounding like a skipping record, we again start with this disclaimer: the definition of mental illness influences both the capacity to identify its prevalence within human populations and also to identify any genes associated with variation in this trait. The modern conception of mental illnesses identifies these as otherwise normal behaviors that, under certain circumstances, become pathological.

A more fruitful way of understanding mental health is to envision it as brain health.[35] The paradigm of brain health makes it easier to understand what makes us vulnerable to mental illness. Just as the other organs of our body are vulnerable, so is the brain. Evolutionary biology has demonstrated that the six factors that make us vulnerable to illness include evolutionary mismatch, infection (viruses/bacteria), evolutionary constraints (limitations in the power of natural selection), trade-offs (all parts of the body have their advantages and disadvantages), reproduction (all of our body systems evolved to promote differential reproductive success), and defensive responses (pain and anxiety can be useful responses to threats).[36] The consequence of the combination of these factors is that by the turn of the twenty-first century, one in five Americans (children and adults) had a diagnosable mental disease as defined by the *Diagnostic and Statistical Manual of Mental Disorders (DSM-IV-TR)* or International Classification of Disease (IDC-10) tools.[37] The major classes of mental illness are mood and anxiety disorders (major depressive disorder, or MDD, bipolar disorder, panic disorder, obsessive compulsive disorder); psychotic disorders (schizophrenia); eating disorders (anorexia nervosa); and childhood disorders (attention deficit/hyperactivity disorder, or ADHD).

Given what we already know about the social determinants of health and health disparity in the United States (see chapter 5), no one should be surprised that this disparity also exists with regard to mental disease.[38] However, the direction in which the disparity exists is surprising. For example, despite the fact that African Americans suffer from more psychological distress, they show lower levels of diagnosed MDD.[39] This conclusion was drawn from comparing measures of psychological distress and MDD in studies spanning 1970–2008, with well over 500,000 subjects. Additional evidence for this paradox is that suicide rates (which are associated with MDD) have always been higher for European Americans.[40]

The rates of post-traumatic stress disorder (PTSD) show no clear patterns of differentiation by socially defined race. PTSD is twice as common in women as it is in men. The environmental risk factors associated with this condition are childhood mistreatment, sexual assault (SA), and intimate partner violence (IPV). The prevalence of PTSD among women who have experienced IPV is six times that of women in the general population. Some studies have found a lower prevalence of PTSD among African Africans and Hispanics compared with European Americans, but those studies were not disaggregated by sex. Others have found the opposite pattern.

National survey data suggest that Africans and Hispanics are more likely than European Americans to be exposed to PTSD risk factors. African American women are more likely to be exposed to IPV, but Hispanic women had a lower rate of assaultive violence than European American women. Other studies have found that European American women were at a greater risk of IPV than African American women.[41] Data from a 2006 study examining the risk of PTSD and MDD in a sample of women included in the National Epidemiological Survey on Alcohol and Related Conditions showed that the risk of a lifetime history of PTSD was 12.79 percent for all women in the sample and 12.68, 13.89, and 12.29 percent for European American, African American, and Hispanic women, respectively. There was no statistically significant difference between the groups. However, for MDD, the lifetime risk was 27.66 percent for all women, and 29.08, 21.78, 24.90 percent for European American, African American, and Hispanic women, respectively. These differences were highly significant. The results are consistent with studies that have routinely found that European Americans are more likely to develop MDD compared with African Americans and Hispanics.[42]

Schizophrenia is a severe and disabling brain disorder characterized by a combination of delusions (erroneous beliefs), hallucinations in every sense (e.g., sight, hearing), disorganized speech, grossly disorganized behavior, negative symptoms (such as flat affect), and, in some cases, catatonic motor behaviors (although the latter might result from the side effects of the drugs used to control the condition). This disease severely impairs social and occupational functioning. Schizophrenia affects one in one hundred Americans, although there is evidence that the rate is higher in African Americans.[43] This higher prevalence is partly influenced by bias among psychiatrists, who are three times more likely to diagnose an African American with schizophrenia compared with a European American patient with the same set of symptoms.[44]

The heritability of schizophrenia is estimated at around 0.80; however, inflammation levels, date of birth, place of birth, seasonal effects, infectious disease (*Toxoplasma gondii* and herpes viruses: cytomegalovirus, Varicella-zoster, and herpes simplex type 1), complications during pregnancy (this actually might be a genetic effect), complications during delivery, substance abuse, and stress contribute to the roughly 0.20 environmental effect.[45] The infectious disease association is also consistent with the inflammation mechanism that influences schizophrenia risk.

Over the last decade, important developments in discovery of the evolutionary origins of schizophrenia have been made. Specifically, we now see schizophrenia and autism as different ends of a spectrum of mental activity.[46] On the psychotic side of the spectrum, the brain produces auditory hallucinations, megalomania, paranoia, delusion, major depression-elation, thought disorder, and mania. On the autism side, the brain produces no or reduced speech, reduced sense of self, lack of mentalistic skill, basic emotions, mechanical logic, and lack of or little goal pursuit.

These disease conditions have their origin in genomic imprinting that results from the fitness conflict between males and females in sexual species. The mechanism of genomic imprinting is epigenetic. DNA methylation can silence or activate genes. Male and female parents imprint genes differently in their offspring to improve their fitness. Male imprinting favors slower growth and larger size at birth, whereas female imprinting upregulates growth genes favoring faster growth and smaller size at birth. This conflict leads to a stable equilibrium in which growth rates and sizes evolve that balance the needs of both sexes. However, mutations can occur that disrupt the equilibrium, and this disruption causes mental disease when the genes involved are expressed in brain tissue.

There is strong theoretical and experimental evidence for this perspective, including genomic experiments with mice and data from human birth cohorts.[47] For example, it has been shown that many of the same genes are associated with autism and schizophrenia; but, more important, a deletion in the gene in question is associated with schizophrenia, and a duplication in the gene in question is associated with autism.[48] Furthermore, the patterns of DNA methylation in genes associated with these conditions differ between persons who have and those who lack them.[49]

Given that the fitness conflicts between males and females and the mechanism of genomic imprinting predates the evolution of our species, it is highly unlikely that specific populations—let alone socially defined races—dramatically differ in the prevalence of autism/schizophrenia spectrum disorder or the genomic variants associated with those conditions. For example, a 2019 study found a common genetic architecture for schizophrenia in Europeans and East Asians.[50] Another study found shared genomic foundations for schizophrenia in Europeans and sub-Saharan Africans.[51]

In summary, envisioning mental illness in the framework of brain health helps us to better understand it as a complex physical trait, similar to intelligence and personality. Furthermore, given the complexity of the human brain, the fact that one in five Americans currently suffers from a diagnosable mental disease in one sense should not be shocking. However, there might be some evidence that evolutionary mismatch is contributing to rising rates of mental illness worldwide.[52] If that is the case, it calls into question ways in which our modern environments differ from those of our ancestors in ways that cause us to be sick. In this vein, it is interesting to note that social injustice has been associated with increased risk of disease (including depression) in a cohort of German workers who were examined between 2005 and 2013. Those who perceived income injustice showed an elevated risk of diabetes, stroke, cardiopathy, asthma, hypertension, and depression.[53] To make the point clear, social injustice makes people sick.

Here, we have not attempted to summarize all the literature on the prevalence of the various mental illnesses observed in the human species. These conditions vary in their heritability, with schizophrenia on the high end (~0.80) and mood and anxiety disorders ranging from 0.20 to 0.60. As with other complex brain-associated traits, the genomic foundations of these disorders are shared across the human species, with populations differing in the frequency of risk-associated variants. Thus, as with other health-related traits, levels of mental health differ in socially defined races, and this variation is not driven by genomic differences. However, the different physical and social environments that socially defined racial groups experience probably play a major role in determining the patterns and prevalence of mental illnesses observed within them, as in the cases of MDD and PTSD.

DO RACES DIFFER IN CRIMINAL BEHAVIOR AND VIOLENCE?

Another enduring myth of race is that criminality and violent behavior run along a skin color gradient from light to dark. The false belief that violence and criminality differ by race is part of the discourse that dark-skinned individuals are more physical and unable to control their physical selves, tending toward aggressive behavior. This aggression could be sexual and also result in homicides and other forms of physical harm. The fear of the sexual aggression of Black men was clearly in the mind of D. W. Griffith when he directed the infamous rape scene in *Birth of a Nation*, his 1915 blockbuster that was shown at the White House. This trope of Black criminality was also a component of J. P. Rushton's life-history theory of racial difference. Based on the shoddiest of data, Rushton asserted that Blacks are innately more prone to criminal behavior than whites and East Asians.[54]

In an earlier time, this myth of criminality was routinely used to segregate Europeans of a "color" other than white, including the Irish, Italians, and Jews. Jews were less physical but more cunning in their criminal behavior. The Irish and Jews, and especially the Italians, became criminal mobsters. During an 1881–82 lecture tour of the United States, Oxford University professor Edward Freeman proposed a simple remedy for what he saw as the social assault against the Teutonic races: "that every Irishman should kill a Negro and be hanged for it."[55] The belief in a higher tendency toward criminal behavior was often supported by junk racist and eugenic science. It was central to efforts to limit immigration by less desirable Europeans, and although the science has been discredited, the ideology remains.

Today, the greatest myths of criminality are used against brown and Black peoples. There are two separate questions we want to answer: (1) is criminality greater in African Americans and Latinos, and (2) if so, why? A belief in the inherent nature of criminality and race would lead to yes answers for the first question. We find that the prevalence of acts defined as criminal might be greater in these groups, but the reasons have nothing to do with biology and a lot to do with racism.

Similar to our discussion of intelligence and personality, we must begin with a definition of *crime*. Legal definitions describe crime as behavior by an act or omission, as defined by statute or common law, deserving of punishment. Sociology takes a broader view, describing crime as deviant behavior that violates prevailing norms concerning how an individual ought to behave. In both cases, we have the problem that depending on the law or social norm in question, an individual acting morally can be considered a criminal. For example, Lutheran theologian Dietrich Bonhoeffer (1906–45) was hanged for his opposition to Nazism, Rosa Parks (1913–2005) was arrested for defying the city of Montgomery's segregation ordinances, and Nelson Mandela (1918–2003) was imprisoned for most of his adult life for opposition to apartheid in South Africa.

On the other hand, clearly immoral acts such as financial crimes are often applauded by society even though they violate the law. *Business Insider* listed the ten biggest financial white-collar crimes from 2000 to 2009. Each of the individuals listed was a man of European descent, including people like Kenneth Lay, who, with the accounting firm Arthur Anderson, mispresented Enron, losing shareholders $1 billion. Lay was never convicted of a crime. Others, such as New York City money manager Bernie Madoff, ran a classic Ponzi scheme for over two decades, defrauding investors of an estimated

$65 billion. He was convicted and sentenced to 150 years in prison.[56] However, whenever racial theories of crime are discussed, they usually focus on violent crime, such as rape, homicide, and serious assault. For example, a review of the 2020 tables of contents in the academic journal *Criminal Justice and Behavior* did not reveal a single article associated with white-collar crime.[57]

What causes one to be a violent criminal? This question is also conflated with the issue of what counts as criminal behavior and what is more properly understood as mental illness. For example, are antisocial behavior, conduct disorder, callous unemotional affect, and antisocial personality disorders predictors of crime? Or are these best understood as mental disorders that might or might not be associated with an individual who commits a violent crime? As with intelligence, there have always been competing hypotheses, with nature on one side and nurture on the other. Are criminals born that way, or are they made so by their life experiences? Those supporting the idea of an inborn tendency to criminality have pointed—often with flawed data—to how crime runs in families. They have used skull shapes (phrenology), hormone levels (Rushton's theory, mentioned earlier), and, recently, genes to detect criminals.

Our broken-record analogy shows that genome-wide association studies of traits possibly linked to criminal behavior do not show any evidence of genomic differentiation between socially defined racial groups. A 2019 review of genes associated with aggressive behavior (antisocial behavior, conduct disorder, callous unemotional affect, and antisocial personality disorders) examined about 59,000 individuals (mostly of European descent, with some African Americans) across six studies published between 2011 and 2017 and found only five statistically significant SNPs for the four traits under study.[58] Those data do not make for a very strong genomic foundation for a criminal behavior argument.

In 1992, with crime increasing in many urban areas in the United States, the National Institutes of Mental Health (NIMH) began its Violence Initiative to study the biology of crime and violence. Black and brown inner-city youth were compared to animals roaming the urban jungle. It was in this ideological climate that the case of the so-called Central Park Five emerged. In this horrific injustice, five Black and brown teenagers were accused of the rape of a white Central Park jogger and convicted by coerced confessions and the slimmest of evidence.[59]

TABLE 7.3
Probability of Homicide by Race of Victim and Race of Offender

	White	Black	Other	Unknown offender
White	**0.81**	0.09	0.02	0.02
Black	0.08	**0.89**	0.01	0.03
Other	0.25	0.18	**0.55**	0.02
Unknown victim	0.42	0.22	0.06	**0.30**

Note: The boldface numbers show the homicide probability within socially defined group. Thus, a white American is nine times more likely to be killed by a white than a Black American, whereas a Black American is ten times more likely to be killed by a Black than a white American.

The data are modified from U.S. Department of Justice, Federal Bureau of Investigation, Criminal Justice Information Services Division, "2018 Crime in the United States," table 6: Expanded Homicide Data, https://ucr.fbi.gov /crime-in-the-u.s/2018/crime-in-the-u.s.-2018/tables/expanded-homicide-data-table-6.xls.

Although the panic over the alleged criminality of Black and brown men is almost entirely about fear of whites being the victims, such occurrences are actually rather rare. What is unfortunately more frequent is that victims of Black and brown crime are most often also Black and brown. For example, homicide in America occurs mainly within socially defined racial categories (table 7.3). In 2018, 3,315 whites and 2,925 Blacks were murdered in the United States. However, whites make up 61 percent in the U.S. population, whereas Blacks account for 13.4 percent. So, to generate equivalent numbers, you would have to divide the white number by 4.55, resulting in about 728 whites murdered for equivalent population sizes. Even more telling is what's revealed when examining deaths due to homicide by race (fig. 7.1). Blacks show the highest mortality rate by homicide across age. For all other groups, there is no detectable homicide from ages one to nineteen. Indeed, the rank of these groups for homicide mortality tracks with their median net wealth per family.[60]

The question is, why? If one does not study in detail the causes of violent crime, one could be left with the impression that the racial variation is innate. It isn't.

Two studies by epidemiologist Brandon Centerwall show how important it is to try to control for environment when comparing racial differences in violent crime. Centerwall, working at the Centers for Disease Control in the 1990s, came across federal data on rates of domestic homicide in different cities. A domestic homicide is the murder of one individual by another

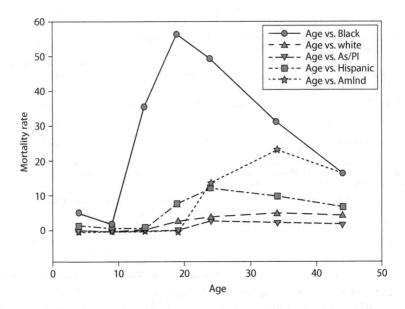

FIGURE 7.1. The homicide rate per 100,000 individuals by socially defined race in 2017. These figures include both sexes. The greatest gap is observed at ages 15–19, when the rate for Blacks is 35.5 and for whites is 2.4, with all other groups at about 0.0. At later ages, American Indians come closest to the rate for Blacks at 23.4 from ages 25–34. *Source:* Melonie Heron, "Deaths: Leading Causes for 2017," *National Vital Statistics Reports* 68, no. 6 (June 24, 2017): 1–77, https://www.cdc.gov/nchs/data/nvsr/nvsr68/nvsr68 _06-508.pdf.

living in the same domestic unit (typically the same apartment). Centerwall first studied intraracial domestic homicides in Atlanta, Georgia, and New Orleans, Louisiana, for a number of years from the late 1970 to the mid-1980s.

Centerwall found that for both cities, the risk of intraracial domestic homicide was almost six times greater in the Black than white communities. Others are aware of these data and have left open the interpretation as to why homicide rates vary so much. This study went further; Centerwall developed a simple measure to control for crowding by stratification of the socioeconomic status of different census tracts. He found that nearly all white residents of these cities lived in areas with low crowding (and higher SES), whereas Blacks often lived in areas with much greater crowding, and the rate of Black domestic homicide increased dramatically with crowding.[61] Furthermore, if one controls for crowding, comparing the rate of Black versus white domestic homicide in census tracts with equal crowding, the rate of homicide is almost equal. Centerwall concluded that domestic homicide has nothing to do with race but a whole lot to do with crowded conditions, which are themselves attributable to racism and class.

The myth of criminality is powerful in how it has burdened Black and brown families. Police are not helpful or friendly; they are more likely to harass, arrest, and even kill. Police encounters lead to incarceration, and incarceration leads to further struggles for families of the incarcerated individual to maintain anything close to a healthy lifestyle. This is systemic racism.

CONCLUSIONS

Individuals vary in complex traits such as musical ability, personality, and intelligence. The cause of these traits is equally complex. Modern genomics has demonstrated that although genes play a role, individual variants are likely to make a very small contribution to complex traits. Rather, genes work together with chance and environmental conditions. After all, these complex traits, including brain diseases and violent criminality, are the result of a cascade of events that might stretch back generations and continue to trigger specific violent events.

There are some measured differences in these complex traits across populations and socially defined races. Where we have failed intellectually is in gravitating too quickly to the simplified assumption that these differences are due to more consequential biogenetic differences. This assumption is unfounded. Indeed, much simpler environmental factors, such as poverty, have been demonstrated as explaining most of the variation in violent crime rates.[62] Indeed, the UN Global Study of Homicide shows a clear inverse relationship between a country's gross domestic product (GDP) and its homicide rate. To underscore the complexity of the association between homicide rate and social, cultural, and environmental factors, some of the countries with the highest GDPs (Germany, Switzerland, Japan, Hong Kong) have the highest proportions of female homicide victims.[63] Simplistic genetic theories of violent crime are scientifically untenable and have done tremendous harm in supporting the status quo and worldwide racial hierarchies.

Chapter Eight

DRIVING WHILE BLACK AND OTHER DEADLY REALITIES OF INSTITUTIONAL AND SYSTEMIC RACISM

The existence of racism is not up for debate. An overwhelming abundance of facts demonstrate the reality of structural racism in America. Unfortunately, we have not reached racial equality. We are not postracial. The truth is in the facts of racial inequalities.

This chapter is about racial inequality in life circumstances and opportunity—today. These inequalities have histories and origins in the past. They have a legacy. They are part of the long and continuing history of institutionalized and systematized unequal treatment by race. Here, at the onset, we remind you that racism isn't about individuals, or thoughts, or specific acts, or even intention. Rather, racism is about systematic maltreatment by groups with access to power over groups without equal access to power. Thoughts and ideas, no matter to what degree they are explicit and implicit, are important because they solidify into ideologies and worldviews that justify institutions and systems. And the consequences of racism result from institutions and systems of racial oppression. This means that addressing one's own racial misconceptions is an important start but very far from the end of eliminating racial injustice. We also need to go beyond the individual awakening, because ending racial injustice requires systemic and concerted political and social action.

The consequences of systemic racism are found pretty much everywhere one looks. One finds, without much digging, that racial inequalities are at every step of medical care and health outcomes (see chapter 5), from general causes including unhealthy diets and air, pollutants, dangerous streets, and lifelong psychological stresses, to lack of health insurance and lower-quality and less aggressive medical care. Racial inequalities are also found in education, beginning with less access to quality childcare and early-childhood enrichment programs, impoverished schools from pre-kindergarten to college, and systemic lower expectations and ability to stay in school. Inequalities are part of the housing and neighborhoods of racially subordinated communities. These environments include heavy metals in soils and chipping paints in buildings, unsafe streets, and houses without proper heating, cooling, and ventilation, all of which are among the various aspects of environmental racism.[1] Racism puts up obstacles to secure, prestigious, and well-paid jobs. Biases become institutional racism when an employment selection process results in choosing applicants and advancing workers whose résumés look white, whose names sound white, and who fit most seamlessly into the white club.

All of these biases and actions contribute to a silent system that leads to inequalities in family wealth. Poor education, health, and environments lead to lower-paying jobs. And the trend extends from the past, as less inherited wealth and building of wealth through homeownership and other long-term investments perpetuates intergenerational cycles of family poverty.[2] The cycle continues, with lower wealth leading to poor education, poor diets, exposure to toxic environments, and poor health.

Inequality of opportunity by race is central to our history. It is intergenerational. It is a cultural inheritance. It is part of the social contract of the United States. Just as in genetics, DNA is not destiny; racial inequalities and racism in societies are not forever, and acting to eliminate them is under our control. In fact, part of what excites us is that social ideas can change. We are a flexible species with brains that adapt. We can and must change. Racial inequality is ours to overcome. The future of our species is at stake.[3] To understand why overcoming racism is so imperative, let's take a look at the depth and breadth of racial inequalities and racism.

In prior chapters, we have defined and discussed racism and its various forms. We have explored in depth how racism, rather than biological race, impacts health and longevity (chapter 5). In this chapter, we examine some of the other consequences of the lived experiences of racially subordinated people in a society with endemic racism. We bring to the forefront some of the many everyday experiences of racism in society. Our focus is on five interrelated arenas where racism lives today: the criminal justice system, education, employment, environments, and wealth accumulation.

DRIVING WHILE BLACK, OR WHY DO AFRICAN AMERICAN MEN GET PULLED OVER BY THE POLICE MORE OFTEN THAN OTHERS?

On July 6, 2016, Philando Castile, a thirty-two-year-old Black man, was driving in a suburb of Saint Paul, Minnesota, when he was pulled over at 9:00 P.M. by police officer Jeronimo Yanez and his partner. Castile was driving with his partner, Diamond Reynold, and her four-year-old daughter. Once stopped, Castile, with his seat belt on, is heard on a video telling the officers that he had a firearm that he was licensed to carry. The video shows Yanez becoming agitated, screaming "Don't pull it out!" Both Castile and Reynold try to assure Yanez that Castile is not pulling out his gun. But the agitated officer shot Castile at close range seven times, hitting him five times and piercing his heart. Castile died at 9:37 P.M. This was Castile's forty-sixth traffic stop.

Driving while Black (DWB) sounds kind of cool and flip. It is a word play on driving while intoxicated (DWI). And it is also equally serious. Driving while Black refers to a part of the system of differential treatment that Black- and brown-skinned individuals, especially Black men, receive from law enforcement. Statistics repeatedly show that Black men are more often stopped and searched by police officers than other men.[4] A study conducted about a decade ago in St. Louis, Missouri, showed that searches were more likely in stops of Black drivers than of white drivers, especially those conducted by white officers; the study controlled for other characteristics of the officer, driver, and reason for the stop. In predominantly Black communities, however, stops of white drivers by white officers were more likely to result in a search. The reason for this was that officers have assumed that a likely reason for white drivers to be in Black communities was to purchase drugs.

A recent study of more than 20 million traffic stops in North Carolina across multiple decades showed a pattern of more aggressive targeting of racially subordinated youth, despite the fact that the stops were not justified by differential crime rates among those youths compared with whites. Finally, to demonstrate that this problem does not reside only in the former Confederate states, the stop-and-frisk policy of New York City police was also shown to be racially biased (controlling for social areas, neighborhoods, and precincts).[5]

Scholars and activists have pointed out that over-policing of racially subordinated people is rarely comprehended by middle-class whites. This might have contributed to the grief and anger many white Americans experienced when they saw the full video of George Floyd's murder on television news and social media feeds. Routinely being pulled over while DWB is part of systematic unequal treatment in the criminal justice system. Black and brown communities are differentially targeted, accused, arrested, found guilty, and incarcerated. Black and brown people receive unequal treatment at every stage in the criminal justice system.[6]

Why? Driving while Black is justified as part of racial profiling, the use of perceived race as grounds for suspecting someone of having committed an offense. However, the problem is that racial profiling is based on a false supposition that Black and brown people are more likely to be breaking the law. For example, the war on drugs, the increased criminalization of not just selling but being caught in possession of small quantities of marijuana and other illegal drugs, targets Black and brown communities.[7]

In *Suspect Citizens*, Frank Baumgartner and co-authors document race inequalities in traffic stops.[8] Hispanic drivers are pulled over about as often as white drivers, but once pulled over and perhaps identified as Hispanic, they are more likely to be searched. Even though whites drive more often on average than Blacks, the authors found that Blacks are nearly twice as frequently pulled over as whites. It doesn't stop there. Blacks, once pulled over, are twice as likely to be subject to search. So, overall, the chance of a Black driver being searched is about four times that of a white driver.

Many might think that this sort of profiling is justified. It is common, but it is not effective. After these traffic stops, police are less likely to find guns, drugs, or illegal substances in the cars driven by African Americans or Hispanics. The contraband hit rates are 36, 33, and 22 percent for whites, Blacks, and Hispanics, respectively. In fact, if anything, the cause-and-effect relationship is reversed. Racial profiling of Black and brown men and women leads to further arrests, convictions, and jailings, as individuals without white skin privileges go through a system of inequalities. The real reason for the driving while Black phenomenon is racism.

Sandra Bland, a twenty-eight-year-old Black woman with a new job, died in her jail cell of an apparent suicide. Three days earlier, July 10, 2015, Bland was stopped for changing lanes by a Texas State Trooper. She said she had changed lanes to let the trooper's car pass. After an exchange with the officer, Bland was handcuffed and jailed. Records indicate that this was Bland's tenth routine traffic stop. Bland, we guess, was done with DWB.

WHY IS THERE A RACIAL DIFFERENCE IN INCARCERATION RATES?

On the warm spring night of April 19, 1989, Trisha Meili was out for a jog in Central Park when she was assaulted and raped. Meili was found early the next morning, so badly injured that she remained in a coma for twelve days. Initially, her identity was hidden from the public. Meili came forward years later to reveal her identity.

This rape took place at a time of heightened fear of Black and brown youth, who were accused of "wilding," a term for running wild in the streets. Some, like Donald Trump, imagined these youth, the Central Park Five and other young boys, as no more than a pack of hunting animals out to destroy property and needlessly harm people. And these packs of youth were said to be Black and brown.[9] Trump publicly railed against the Central Park Five. He paid $85,000 to run ads in New York's four premier daily newspapers calling for the reintroduction of the death penalty. He said, "I want to hate [them]. They should be forced to suffer. . . . I am looking to punish them." All this was said even before a trial date had been set.

Worse, of course, is that these underage men never committed this crime. Feeling pressure to solve the so-called Central Park Jogger case, the police arrested ten suspects. Ultimately, five Black and brown youth, ages fourteen to sixteen, were illegally interrogated and eventually tried, found guilty, and sentenced to five to ten years for the four boys under age fifteen and five to fifteen years for Korey Wise, then sixteen years old. Wise was classified as an adult due to the violent nature of the crime, one we later learned he never committed. He served thirteen years. All the boys had been framed by the New York Police Department, sentenced, and jailed before eventually being cleared of the crime after the true rapist, a convicted murderer and serial rapist, came forward in 2001 and confessed in 2002.[10] His DNA matched DNA found in semen on Meili. He said he had committed the crime alone. Other evidence corroborated his confession.

The case of the Central Park Five shows the logical conclusion of the trope of Black and brown men as bestial rapists and criminals. In the 1990s, the National Institutes of Mental Health (NIMH) focused on aggressive, violent criminal behavior as arising from abnormally high hormones and genetic factors. In 1992, NIMH was beginning to formalize a "violence initiative," except that Frederic Goodwin, a psychiatrist and head of the Alcohol, Drug Abuse and Mental Health Administration, made very clear what such an initiative was about.[11] At a NIMH advisory committee meeting, Goodwin discussed the parallels, as he saw them, between violent behavior by young male rhesus monkeys and young men. He suggested that this parallel might provide insight into the increasing level of violence in inner cities. Goodwin noted that this rise in violence might be the result of a loss of civilizing social factors that typically keep human and monkey aggression in check: "That is the natural way of it for males, to knock each other off . . . maybe it isn't just careless use of the word when people call certain areas of certain cities jungles."[12]

Of course, this sort of reasoning betrays the most simple-minded understanding of violence as part of primate behavior. Social behavior drove the evolution of intelligence in primates. Violence, or its threat, similarly plays a role in all primate behavior and is one of many tactics that can be deployed in the pursuit of social goals. Given that all primates, including humans, are social strategists, they evaluate the costs and benefits of any tactic for obtaining a desired social outcome. Thus, episodes of violence do not result from an individual's "losing control" or from the "beast within" overwhelming the fragile control of civilization. Rather, it is a tactic that individuals rarely choose to deploy because of the very high costs and lower efficiency of other tactics to achieve social goals.[13] This further suggests that in situations of uneven social power, the costs of violent acts might be nil.

This argument would explain the large number of lynchings conducted against Blacks during the era of Jim Crow laws,[14] or the recent wave of shootings of Blacks by white police officers.[15] Alternatively, innocent black and brown men have been jailed or executed by the state for alleged rapes of white women throughout American history. Conversely, a white man has never received the sentence of execution for the rape of black woman in all of America's history.[16]

Racial differences in incarceration follow from racial disparities in all stages in law enforcement and criminal justice, from differences in traffic stops to searches, arrests, convictions, and sentencing. Racial profiling happens on highways and at traffic lights and also in homes, parks, and neighborhoods. Incarceration rate differences are the end result of the system of unequal treatment. But the harm continues for the incarcerated individual and their families, especially their children.

The United States has the largest prison population in the world and also the higher per-capita incarceration rate.[17] In 2016, 2.2 million individuals were in U.S. jails and prisons. As Michelle Alexander aptly puts it, imprisonment is the new Jim Crow. It was not always this way. In 1970, only about

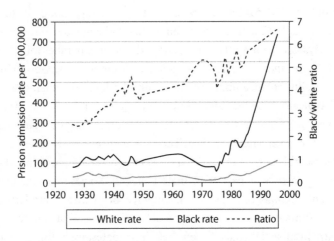

FIGURE 8.1. The rise in rates of prison admission in white and Black populations and the rise in the Black/white ratio. In the 1930s, prison admissions were relatively low, and Black admissions were three times the white admissions rate. By 2000, the Black admissions rate had skyrocketed, especially after 1980, and is now over six times the rate of white admissions. *Source*: Makheru Bradley, "America's Jails Are the Foundation of the Prison Industrial Complex," July 9, 2015, Makheru Speaks (blog), http://makheruspeaks.blogspot.com/2015/07/americas-jails-arethe-foundation-of.html.

400,000 individuals were incarcerated. As shown in figure 8.1, rates began to increase in 1971 with Nixon's war on drugs. Then it skyrocketed further in 1984 with the Sentencing Reform Act. From 1970 to 2016, the number of individuals in prisons and jails grew more than fivefold.

In 1970, the U.S. prison population was not particularly high compared with other countries, especially other Western countries. Now it is. No other country on the planet has a higher rate or number of incarcerated individuals. The rates of incarceration in Cuba, Brazil, and Russia are among the highest but only about half to three quarters of the U.S. rate. Prisons in Europe have about a fifth to less than a tenth the number of inmates compared with the United States.

Why are American incarceration rates so high? The traditional answer is that incarceration is an appropriate punishment for a crime. Slogans such as "let's get tough on crime" are common, uttered by politicians such as Trump and many of his predecessors, and not just Republicans. Getting tough is thought to take criminals off the streets and protects other citizens.

In 1988, Michael Dukakis lost his bid for the presidency to George H. W. Bush largely because of ads suggesting that he was soft on crime. Dukakis was pilloried by a series of ads revolving around a violent crime committed by Willie Horton, a Black inmate on furlough.[18] The Massachusetts Correctional Reform Act (MCRA) of 1972 (which passed due to the leadership of the Republican governor) allowed an inmate who had served more than ten years for first-degree murder to have eighteen to forty-eight hours of unsupervised furlough. During one of his furloughs, Horton broke into the home of Cliff and Angela Barnes, terrorizing and robbing the couple and raping Angela. The ad featuring Horton held Dukakis responsible for allowing this crime to happen because he supported the MCRA.

Many viewed this ad as having played a crucial role in Bush's landslide victory. Although we cannot be certain how much this ad or the public perception of Dukakis's views on crime changed the outcome of that election, clearly, it sounded a warning to American politicians. Now Republicans and Democrats are sure to be seen as tough on crime. William Clinton, the next Democratic president, as well as Hilary Clinton, his wife and future Democratic presidential candidate, established their stances as being tough on crime. Hilary Clinton played on fears by referring to "super predators." Seen as being soft on crime is a sure way to kill a politician's career.

The hope is that incarceration leads to rehabilitation. Unfortunately, that is well known to be a myth.[19] Imprisonment more often leads to a life of crime. It is hard—a small miracle—to escape the cycle because imprisonment leads to impoverishment and a severe restriction of opportunities for a decent, law-abiding life.

Incarceration is big business. There is money to be made in the privatization of prisons.[20] For example, in 2011, corruption at a private prison in Arizona led to a prisoner escape and the murder of a family in Oklahoma. However, because the private prison system is profitable, events such as that did not slow its rise. But there are also economic costs to federal, state, and local governments to incarcerate prisoners. In 2015, the average cost was about $33,000 per year, all paid for by taxes and individual citizens sending financial support to their incarcerated family members. Whether prisons are privately or publicly funded, ultimately, we taxpayers pay for incarceration.

That is part of the problem of incarceration. Less well understood are the costs of conviction and imprisonment borne by both those who are sent to prison and their families. Once a person has a criminal record, it is harder to get a decent job. That much should be obvious. However, what isn't obvious is that socially defined race is a determinant of how much a prison sentence impacts one's employability. One study found that white males with a prison record were more likely to receive a callback for an employment opportunity than Black males without a prison record. This study utilized equivalent résumés, so the Black and white individuals could not be differentiated by any features other than their socially defined race.[21]

Once one enters prison, it is much harder to support a family and provide for dependents. Thus, it also should not be surprising how much the children of incarcerated parents suffer from poorer education and health. About ten million children in the United States have a parent who has been incarcerated. The chance of having an incarcerated parent is twice as high for Hispanic children and six times higher for Black children.[22] This institutional violence is largely hidden from us.

And who goes to prison? Poor people, whites, and especially Black and brown people. Data for 1978–2014 from the Sentencing Project show that the rate of incarceration is over five times greater for Blacks versus whites.[23] As shown in figure 8.2, Hispanics and especially Blacks are disproportionately imprisoned. Of the total male prison population in 2018, 33 percent were Black, 30 percent were white, and 23 percent were Hispanic.

Bryan Stevenson and Michelle Alexander write that imprisonment today is nothing more than the continuation of Jim Crow and slavery.[24] More Black men are in prison now than were slaves in the 1850s! Of course, this is partly due to the growth of the U.S. population, but that fact is still alarming. Today, an astonishing one in three Black men will spend some time behind bars. Such a rate of incarceration is facilitated by an increasing militarization of the police and the differential patrolling of this force in Black and brown

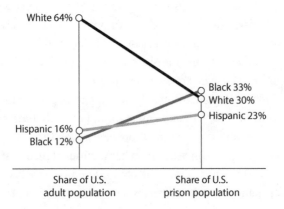

FIGURE 8.2. Share of the U.S. adult population versus share of the U.S. prison population by race and Hispanic origin. *Source*: U.S. Bureau of the Census, Bureau of Justice Statistics.

communities.[25] It is a way to exert control. It is not about safety. It is not about rehabilitation. It is about control. It is about racism.

When criminal (in)justice ends in prisons and jails, the consequences continue in the lives of the former inmates, their families, and their communities. That is why the criminal justice system is the clearest example of institutional racism. But, as we shall soon discuss, it is not the only system.

WHY IS THERE A RACIAL DIFFERENCE IN EDUCATIONAL ATTAINMENT?

First, there is a huge difference in level of education by race. Here, we are talking not about quality of education but simply about final grade completed by age 25.[26] As shown in figure 8.3 from from the National Center for Education Statistics, between 2010 and 2016, 8 percent of whites do not have a high school degree and 35 percent have a college degree. By comparison,

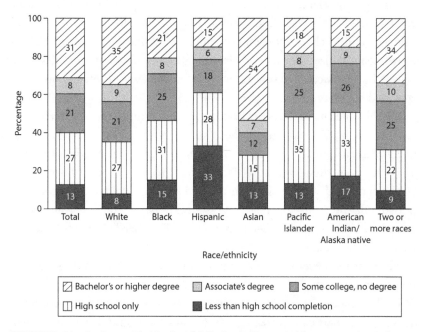

FIGURE 8.3. Education by race/ethnicity. Overall (*left column*), 40 percent of those over age twenty-five in the United States have no college experience, and 60 percent have some experience. Whites and Asians are most likely to have high school experience (65 and 72 percent, respectively), whereas Blacks, Hispanics, Pacific Islanders, and American Indians/Alaska Natives are more likely to have just a high school degree or less education. *Source*: National Center for Education Statistics, https://nces.ed.gov/programs/raceindicators /indicator_RFA.asp.

nearly twice as many Blacks and American Indians and four times as many Hispanics do not have a high school degree. Blacks are 50 percent more likely to not have a college degree, and brown and American Indians are twice as likely to not have a college degree.

The racial difference in education is due to a complex set of causes, all part of the web of racism and socioeconomic inequalities. Clearly, educational attainment is related to socioeconomic class. But it is also related to culture, politics, and racism. Let's break it down.

Education and Social Class

Public school systems are largely funded by local taxes, mainly from real estate taxes, which are related to home value. Wealthier communities with more expensive houses pay higher taxes, which they are able to afford (in part because their houses increase in value, thus building wealth), and with these extra funds they are able to spend more for better schools, more teachers, and more educational resources. They can afford special education teachers, teachers with advanced degrees and enthusiasm for their subjects and students, and teachers with resources including books, computers, musical instruments, and science labs. Their students have more opportunities to be engaged with educational subjects in ways that are exciting, often individualized, and hands-on.

On the other hand, students from poor communities go to underfunded schools. We have already established that the poverty rate of Black, brown, and American Indian families is much greater than that of white families. The schools in these communities are under-resourced and under-stimulated. School and learning are to be avoided rather than enjoyed. Schools are not stimulating. Today, increasingly, as wealth varies, students with means can afford better schools, whereas poorer families are left with less. Students from wealthier families can attend special charter and private schools. If higher education ever was an institution that allowed for social mobility, it no longer does so.

Public schools in poorer communities typically have fewer resources. This might be particularly true for schools on American Indian reservations. Because white families are less likely to live in such places, poor white children are less likely to attend these schools.

Alan's daughter Ruby was fortunate to go to elementary school in Leverett, a small rural community that is part of the Amherst, Massachusetts, school system. Amherst is a liberal town with many families associated with the University of Massachusetts (UMass), by far the largest employer. Leverett Elementary School was built for more than one hundred fifty students but currently has only about one hundred, almost entirely white students. Opportunities are abundant for learning and getting individualized attention. Similarly, Joe's children (Joey and Xavier) attended elementary school in an upscale neighborhood in Glendale, Arizona. Most of the students' parents held professional jobs. An outstanding feature of the school was the fact that several college-educated moms were available to provide in-classroom support to teachers.

Ruby was having trouble hearing and retaining instructions. One of her teachers had an idea. Perhaps her issue was with central auditory processing. The school district made plans for Ruby to be tested at a UMass laboratory. And, yes, that was the problem. Next, they had her work with a speech pathologist and purchased an expensive computer program. Ruby's auditory processing improved dramatically. She was lucky. But her luck is also part of an unequal system.

In addition, the double whammy of being poor and nonwhite results in teachers' having low expectations. One of the biggest myths of race (see chapter 7) is that there is a racial (and class) difference in intelligence. Joe began school in 1960 in an integrated school district in Westfield, New Jersey. At that time, students were tracked into slow, intermediate, and advanced learning classes. Joe began kindergarten in a slow class (all the Black kids in the school were assigned to this class). As a young child who was not challenged by his work, he soon was identified as a discipline problem. The school organized a parent-teacher conference and recommended to his mother that Joe be placed in a remedial class. "Remedial," at the time, was a code word for mental retardation or learning disability. Joe's mother, who had been raised in the Jim Crow South, would not allow it. Three years later, a student teacher found Joe in the library reading *The Crusades* by Harold Lamb.[27] Lamb's book was used in college-level history courses. The rest of the teachers had assumed that Joe was pretending to read such books due to his inability to read. However, this student teacher was bold enough to sit down next to him and ask him about what he was reading. Joe then proceeded to explain the main points of the book, his favorite events, and historical characters. The next day, Joe was reassigned to the advanced classes, the only Black kid to be so placed. The rest is history. Unfortunately, we have no idea how many other Joes were never discovered. That is a loss for our society and an especially painful loss for the undiscovered Joes.

The Pseudoscience of Low Expectations

In 1967, Arthur Jensen began a long, widely read, and attention-grabbing article in the *Harvard Educational Review* with this sentence: "Compensatory education has been tried and it apparently has failed." What Jensen meant by compensatory education were efforts such as the Head Start program, intended to provide resources for children of poorer families and communities. He cites Head Start's own statistics showing that an initial gain in IQ test scores of five to ten points in children from these programs tended to diminish over time.[28]

Was that a failure? We think not. First, a gain in five to ten IQ points for just a little enrichment is astonishing. The failing is in considering that a little educational enrichment for a short time could overcome a life of unequal treatment—in home environments, diets, medical care, and educational resources, among other areas.

Joe has personal experience of this. In the mid-1990s, he helped to direct a National Science Foundation–funded project, the Phoenix Urban Systemic Initiative. Its goal was to provide greater opportunities for racially subordinated urban youth to enter the science career trajectory. In one of the poorest neighborhoods of the city, an elementary school with primarily Mexican American enrollment was performing as well as or better than the other schools in the district. Many of the children were from families headed by a single mother. The secret to their success was that the school was located in the shadow of Sky Harbor Airport and was receiving funding from taxes paid by the airport. As a result, this school was abundantly provisioned with teachers, supplies, and other resources. However, after these children left that elementary school, they were funneled into middle and high schools without that sort of support. Without great surprise, their test scores began to fall, such that their scores were equivalent to those observed for other poor Mexican Americans in the Phoenix School District.

Despite the glaring examples of "savage inequalities" in school funding, it is convenient to blame unequal education and educational attainment on genes and racial differences.[29] Jensen's article echoed prior pronouncements going back to Samuel Morton's study of cranial capacities and going forward to Richard Herrnstein and Charles Murray's *The Bell Curve* in 1994 and Nicholas Wade's ideas that societal successes are based on genes and race.[30]

No, gaps in educational achievement have simple, systemic explanations. Wealth is a determinant of educational achievement, and educational achievement leads to wealth. Low expectations, a part of systemic racism, lead to low educational achievement. They are parts of a system that reinforces and reproduces itself throughout history.

As we were writing this chapter, then-President Trump was telling all who would listen that the coronavirus will just go away soon. Just like that. Just as we know that was a lie, it is misleading and incorrect to think that educational gaps by race and class are due to genes. We discussed the poor science behind genetic and racial claims of intelligence and educational attainment in chapter 7. In the case of both the COVID-19 pandemic and inequalities in education, action to address them is sidelined. But the truth is that educational inequalities are fed by inequities in environments and expectations that feed employment and income inequities, and all of these systems of inequalities work together.

DOES PERCEIVED RACE AFFECT EMPLOYMENT OPPORTUNITIES?

Yes. There are two parts to consider, the immediate and the systemic. The immediate is evidence that individuals with Black-sounding names are far less likely to have their résumés taken seriously. This impact has been documented over and over again. For example, Bertrand and Mullainthan sent out fictitious résumés in response to help wanted ads in Boston and Chicago.[31] They randomly gave the résumés, which were otherwise identical, African American or white-sounding names: Lakisha and Jamal versus Emily and Greg. White names received 60 percent more callbacks for interviews. The gap is unaffected by occupation, employer size, and industry. It is universal.

Although it is studied a bit less often, the same is true of Latinx- and Asian-sounding names. Common practice today, especially in the Asian community, is to "whiten" résumés by adding white-sounding hobbies and making subtle changes in names, for example, from Lee to Luke.

The longer-term, systemic aspect relates to the "old boys clubs" of corporations. Class and gender can intersect with race. As a long-time faculty member, Alan can attest to the number of times an applicant has been deemed comfortable and thought to "fit in" better. Joe lost a lot of faculty appointment opportunities early in his career because he had the reputation of being "too Black." The applicants who make predominantly white faculty members comfortable tend to be from the upper class, have come from elite schools, and, most notably, are white. This injustice persists for science faculty, as discussed by Joe and Erich Jarvis.[32]

A good part of the reason that African Americans and Latinx are over-represented in some types of employment is that they are steered away from stable, well-paying, and high-status jobs. For decades, it was difficult for non-whites to get a job in a union or a solid blue-collar job as a policeperson or firefighter. Federal laws opened up these jobs, slowly. However, over the latter part of the twentieth century, the American labor force began to polarize, with high-skilled occupations that require many years of education at one end and low-skilled, precarious employment at the other. Due to inequities in educational opportunity, Black and brown people have been differentially regulated into the precarious workforce.[33] This is also an example of a color-blind racial inequity. On the surface, it appears as if this pattern results from blind forces of the market, but in reality, it has been created by the structural racism of the American educational system.

WHAT IS ENVIRONMENTAL RACISM?

Environmental racism refers to the differential exposure to pollution and unsafe environments by race. Poor communities and communities of color are most likely to be near polluted areas and exposed to chemicals, such as lead, that might stunt development and cause health problems such as asthma.[34]

Like a lot of undergraduates, Joseph Jones was not sure what he wanted to study at Howard University. The son of a minister with lots of enthusiasm for ideas, Joseph had a whole host of intellectual interests. He loved history, especially African American history, and philosophy, sociology, and the sciences. He thrived at Howard, taking courses in a variety of departments but without a singular direction. Then, one day, he took Introduction to Physical Anthropology with Professor Michael Blakey. Blakey is a leading African American anthropologist and perhaps the best-known African American anthropologist working on race and slavery. Jones was invited to see if he might like to work in Blakey's lab, named after famous African American scientist W. Montagu Cobb, looking at the bones of enslaved Africans from lower Manhattan. Jones was hooked on the scientific study of bones to tell something about the hidden history of enslaved Africans.

Years later, as a graduate student, Jones worked with Alan at the University of Massachusetts. His main goal for his dissertation was to try to figure out who was born into slavery in New York and the American colonies and who might have grown up in Africa, became enslaved, made the middle passage, and came to New York as an adolescent or adult. His novel method to answer this question was to assess the unique chemistry of growth rings in teeth that calcified in early life. Different landscapes lead to different chemical signatures in these tooth rings. For example, West Africa has a ratio of strontium isotopes different from those of the eastern shores of North America.

Jones was able to find these signals and separate who had been born enslaved from those who became enslaved after their birth in Africa. He also found that the lead levels in the teeth of those born in Africa were low, not at all surprising for individuals from the eighteenth century. And then came the surprise: the lead levels of those born in the Americas were high, even in their first years of life. It was often off the charts.

Now, three hundred years later, lead pollution is still getting into poor bodies, and those poor bodies are often from poor communities of color.

Lead has long been known to be a neurotoxin. In any amount, it affects neurological development. There is no safe amount. And the more lead, the worse and more irreversible the effects. Despite early suspicions about the toxicity of lead, from the 1920s to 1973, it was added to gasoline in the United States (to prevent engine knocking) and to paints that went on houses, windows, and almost everywhere else. Lead was not phased out of gasoline until 1986. In many of the houses painted with lead, coats of lead paint are still chipping and peeling. Lead is now in soils. Lead also remains in water pipes, such as in Flint, Michigan, and it continues to be found in the bones of pregnant women and released into their fetuses.

A recent study compared blood lead levels of young Black, white, and Hispanic children based on a national sample from 1988–91, 1991–94, 1999–2002, and 2010. In 1988–91, around 18 percent of Black, 7 percent of brown, and 6 percent of white children ages one to five years were shown to have at least 10 µg/dL (micrograms per deciliter) of lead in their blood, a criminally toxic level. Black children were three times more likely to have high blood lead levels than white children.

The overall levels have dropped from the time of that study, but in 2010, young Black children were still 2.8 times as likely as non-Black children to have elevated blood lead levels, after adjusting for other risk factors.[35] The source of the lead exposure could not be ascertained. However, it is likely to not be a single source, possibly from the bones of mothers, soil, paint in old buildings that have not had lead paint removed, and pipes such as in Flint and other cities with unattended infrastructures. Lead pollution is silent, yet it is a potent example of environmental racism.

There are a number of aspects and forms of environmental racism. In general, poor communities of color are more crowded. Apartments and other types of housing might have less ventilation. Communities also might be food "deserts," places where it is difficult to find high-quality foods. These places also might be where it is easier to buy liquor than a fresh apple. It is easier to get a cheap taco, soft drink, or burger than a peach or tomato.

Poor communities are also safety deserts. Residents lack places to hang out. These communities might even lack shade trees, which are taken for granted in white communities. The trees provide protection from the sun and cooling.

As industrialization has increased, factories have released pollutants into the air, water, and soil. These factories are most often located near poor communities. Environmental racism is racism because it is the result of a difference in power. Wealthy and white communities have the political power to resist factories. They have a tax base of sufficient size that they do not need industrialization.

IS THERE A RACIAL DIMENSION TO CLIMATE CHANGE?

Yes. A crisis, whatever it might be, is almost certainly going to be amplified in poor and nonwhite communities. And climate change is bringing on crises.

Liberal whites and many others often think that ecology and the existential threat of climate crisis are not as important to Black and brown as to white communities. The environmental movement, until very recently, is led largely by liberal whites. And, indeed, poorer communities face other threats that compete with threats to ecological conditions and global warming. However, poorer communities are also the most vulnerable to and will be most impacted by climate change.

One need only to look back to Hurricane Katrina in August 2005. We will not get deeply into how much Katrina was aggravated by global warming. The degree that one attributes anthropogenic climate change with causing the greater frequency and intensity of severe weather is inconsequential, even though, at the time of this writing, two hurricane-force storms— Laura, at category 4, and Marco—were bearing down on the Gulf Coast.[36] We are concerned here with the response and human costs of Hurricane Katrina to the residents of New Orleans.

Katrina's storm surge caused as many as fifty-three flood protection structures in and around New Orleans to fail. The result was that 80 percent of the city was under water. The total damage was estimated at $125 billion. The human toll in suffering and loss of life is harder to pin down.[37]

Quite simply, most New Orleans residents in wealthier, predominantly white communities were able to safely escape and ultimately return quickly to their pre-Katrina lives. On the other hand, residents of nearly the entire lower Ninth Ward were stranded, many lost their homes, and many lost loved ones. The problem is both where individuals live—the conditions under which they live—and how federal and state agencies respond.

On the other end of the scale are wealthy (predominantly white) individuals who choose to live in dangerous environments, such as the coastal scrub of California. These locations have beautiful views of the Pacific Ocean. However, during the Santa Ana winds or lightning storms in the summer, these ecosystems are susceptible to fires. At the time of this writing, wildfires have burned 1.25 million acres in California.[38] Few people recognize these natural disasters as yet another example of white and class privilege. The decision to expand housing into these dangerous ecosystems is driven not by any dire need for new housing but, rather, the power of real estate developers and the market for luxury homes. With the expansion of populations in this fragile ecosystem, water resources are strained, leading to the increase in burnable plant material. This continues to occur despite the well-established scientific knowledge concerning the historic combustion of these ecosystems.[39] The cost of fighting these fires is not shouldered by the wealthy alone but also by the tax dollars of the poor, who do not benefit in any way from the luxury homes of the rich.

The disparity between the politically powerless victims of Katrina and the politically powerful victims of California coastal wildfires is not unusual. Globally, the most dangerous environments are usually occupied by the poor. For example, in countries where poverty is associated with socially defined race or caste, you find disproportionate numbers of these individuals living in low-lying communities that are most vulnerable to hurricanes, rising tides, and infectious disease. Alternatively, wealth allows people the opportunity to choose to live in beautiful but often dangerous ecosystems. The capacity to do so is often subsidized at public expense, thus doubly victimizing poor people.

WHY IS THERE A RACIAL DIFFERENCE IN FAMILY WEALTH?

Here's the short answer: history. A history of racism and unequal treatment now has consequences for the accumulated wealth of families of different social races.

The gap in wealth between white families and Black, brown, and American Indian families is huge and hugely significant. The example of how the Five "Civilized Tribes" (chapter 5) were dispossessed of their farms and livestock and forced to take a hazardous journey into inhospitable lands west of the Mississippi River demonstrates how wealth is influenced by a history of subordination. Contrary to most people's understanding, the federal government has done far more to help white people gain wealth than it ever has for nonwhites.[40]

Wealth is the total amount of monetary assets one owns minus all that one owes, or debt. Alan and Joe differ in social race but share class backgrounds. Their current wealth includes the value of their homes (minus what they owe in mortgages) and all of their other assets, such as retirement accounts minus any other debts like car loans, credit card debt, and student loans. Both of us are one generation out of the working class. This means that we had debt (college loans) to cover. Alan paid his off when he became a full professor. Joe has still not finished paying his loans. They cannot pass on an inheritance in the same way that a person who is descended from the middle class (professionals, doctors, lawyers) or upper class (business owners) can pass on to their children.

By comparison, income or annual income is the amount of money one gets in salaries, wages, investments, and the like. Income obviously can make wealth, especially if one has income above what is needed to live, which can be invested. As well, individuals in the top income brackets often earn income not just from their jobs (employment income) but also from the return on their investments. Money makes money.

Wealth is more important than income, because it is wealth that one can use to invest in the future. Or wealth can be passed on, inherited.

Wealth is sort of like mortality, the end result of a process that likely involves one or more diseases. Wealth is where we see the most glaring racial disparities.

As we write, the average wealth of a Black or brown family is decreasing. In 2016, white families had median wealth (that is, half had greater and half had less) of $171,000, whereas Hispanic families had median wealth of $20,700 and Black families had median wealth of $17,600, or about a dime for every white dollar.[41]

White families are more likely to own their homes, to have accumulated wealth in homeownership through appreciation, and to own businesses, stocks, and retirement accounts. The wealth disparity has huge implications for how race is lived in America. It determines access to education, comfortable and safe housing, and health care. This in turn determines how long a person will live. Indeed, the disparities we are currently observing from COVID-19 infection and mortality are strongly correlated with wealth.

CONCLUSIONS

There are only two ways to explain all of these racial differences in health, incarceration, educational achievement, employment, and wealth. One can explain that they are all due to genetic differences, or one can explain that they are due to systemic inequalities. Throughout this book, we have shown that the first explanation—genetic differences—lacks any support. None. Zip. So, the only explanation is that the differences are due to systemic inequalities. We hope that is sobering, clarifying, and ultimately liberating.

The truth is plain. We once had a written racial contract that has led to differential access to resources by race for 400 years. African Americans who came to the United States as enslaved persons starting in 1619 have had it the worst. They were enslaved for almost two hundred fifty years and since then have been subject to racist ideologies and systems, such as incarceration and voter suppression, that have kept them relegated to the lowest caste. (Remember the calendar in chapter 5.) Those facts are clear in all the sociological statistics that have been gathered by the nation's finest scholars with appointments at our finest universities.

American Indians were decimated by contact and colonization. They are still recovering from the loss of lands and self-governance. Although the people of these nations often are admired in white imaginations, they nonetheless continue to suffer, often far from the eyes and experiences of the white majority.

Latinx groups have come to the United States at different times and from different areas. But remember that for most Mexican Americans, the border crossed them, not the other way around. The war with Mexico was a blatant act of manifest destiny fueled by southern slaveholders who wanted to expand territory for their peculiar institution. Some Latinx have become "honorary" whites, particularly those light-skinned individuals who immigrated from countries whose political systems our government disfavors. But most suffer as brown people in brown communities, sometimes as much as African Americans.

Asians also have come to America at different times and from different areas. South Asians with darker complexions (e.g., from India) face racism that is similar to that of African Americans.[42] East Asians began arriving in the United States in the nineteenth century (before many white Americans). All faced racist immigration policies and the violation of their rights to become citizens under laws favoring whites.[43] Yet, they have been possibly the most successful in keeping out of the direct line of racist fire. They have achieved a decent level of education. But they are far from free of racism. Witness the way that Trump insisted on calling COVID-19 the "Chinese flu."

What is important to come away with is that racism is a system that intersects different institutions. The exact way the intersection takes place differs by time, place, and racial group. But it is always a system that perpetuates racial inequalities.

Chapter Nine

DNA AND ANCESTRY TESTING

Advances in genomic science have allowed us to conduct a finer examination of the human genome and its variations. We have already established that geographically based genetic variation exists within the human species. We have also shown why this variation is nonracial. Most variation occurs within continents and smaller regions and populations rather than among continents or social races. However, signatures of one's geographical ancestry have been found within our genomes. This might seem like a paradox, but it is not. Ancestry is real. Biological races are not.

In this chapter, we explore why genetic ancestry and biological race are not the same thing. We also examine the methods and claims of DNA ancestry-testing companies. We conclude by putting into perspective the results of ancient DNA research that shows how much humans mixed and mated in the past. These results further clarify why there are not—and never have been—biological races within anatomically modern humans.

WHAT IS ANCESTRY?

Ancestry refers to an individual's history that connects them to past individuals, families, traditions, and places. Ancestry is a concept that is used in popular discourse and by many cultures and religions to trace individual and family connections. Many of us build family trees and now take DNA "ancestry" tests to help us figure out one type of ancestry that can be traced through genomes. We humans have a yearning to make these connections.

We find it useful to think about three types of ancestry: cultural, geographic, and genetic. *Cultural ancestry* refers to the linguistic, social, cultural, and religious traditions that one follows and is connected to. These might include dialects, ways of preparing food, residential patterns, and beliefs that both connect us to the past and changes with the times. Culture does not sit still. Cultures and components of cultures can fade, merge, and grow. Cultures evolve and change.

Geographic ancestry refers to where one's ancestors came from. This might be known through family histories, as well as through written records such as shipping information and archaeological evidence such as artifact similarities. Recent advances in paleo-geochemistry have made it possible to trace ancient migrations and the movements of individuals and groups by matching the chemistry of early-forming bones and teeth to water and soil. For example, in a study of enslaved individuals from the New York African Burial Ground, Alan and his colleagues were able to determine which individuals were born and spent their infancies in the Americas and those who were born in Africa and the Caribbean.

What these studies cannot tell is where one's ancestors grew up. This information can come from family histories and sometimes written sources. Alan's grandparents came to the United States sometime in the early 1900s from somewhere in Russia or Poland, escaping pogroms and anti-Semitism. His family history pretty much stops with his parents. That's probably not unusual. Geographic ancestry is complicated, because we humans tend to move around and often do not think to pass the information down the generations.

We most often think of the third type, *genetic ancestry*. Genetic ancestry is written in DNA and passed through both maternal and paternal lines. Getting information about one's genetic ancestry, currently referred to as finding one's ancestry or roots, has become extremely popular with the advent of inexpensive sequencing technologies. For example, African Americans can now get information about the group in Africa from which they might have come.

Finally, it is important, when thinking about these types of ancestries, to recognize that they might or might not be related. Geographic, cultural, and genetic ancestries are often confused with one another. In the earlier example, individuals born in what today is Ghana and enslaved as adolescents in the eighteenth century spent their adult lives as slaves in America. Their geographic and cultural ancestries shifted, but their genetic ancestries did not. Similarly, individuals fleeing poverty in Ireland or China in the nineteenth century came to Boston or San Francisco, and their genetic ancestries tie them to their natal homes. Their bones and teeth also have telltale signs of their natal homes, but their culture slowly shifted to Irish American and Chinese American. A child adopted from Korea by European American parents grows up in their parent's Euro-American culture, but that child is genetically and geographically connected to East Asians.

IF RACE ISN'T GENETIC, HOW COME MY DNA TEST RESULTS ARE TELLING ME MY ANCESTRY?

As of February 11, 2019, more than twenty-six million people had taken a direct-to-consumer (DTC) genetic ancestry test.[1] As we write, at least thirty-eight companies offer such tests. Many of these companies employ the concept of continental ancestry in reporting ancestry results to their customers. This concept assumes the existence of a small number (typically four or five) of ancient parental populations that gave rise to our current populations. It is frequently equated with race in the biological sense. On the other hand, biogeographic ancestry is the association of a person's origin with the geographic location(s) of presumed ancestors. This is inferred by comparing the relative similarities of an individual's genetic markers with those found in contemporary populations.

To understand how one can make this inference, it is important to understand the nature of the genetic markers being used. We have already pointed out that only about 1.5 percent of the human genome is composed of protein-coding sequences. It has been estimated that the total amount of human DNA that is exposed to natural selection is less than 10 percent.[2] This means that variation in the remaining DNA is effectively neutral. Thus, that portion of DNA can accumulate mutations, and the specific mutations will be associated with the population in which they first appeared. Of the polymorphic genetic variation within humans, 85–95 percent is found in all world populations. That leaves variation that is associated with continental origin at about 10 percent on the high end and 5 percent on the low end. This means that the variation within subregions of continents is less than 5 percent.[3]

Polymorphism refers to those portions of the genome that can have alternative alleles without causing a major loss in evolutionary fitness. Most loci within our genomes (~67 percent) are monomorphic. This means that a SNP in this position would likely result in death or sterility. Given that the human genome contains about 3.3 billion nucleotides, and that all humans share 99.9 percent of their DNA, the 0.1 percent that is differentiated provides at most 4-5 million genetic markers that could be differentiated by continent and region.

However, for a marker to be useful, its frequency must be strongly differentiated between regions in question, so the amount used by specific DTC DNA companies is less than this amount. For example, the 2015 Bryc et al. study, which examined continental ancestry in Americans, used around 2,020,000 SNPs.[4] The same panel of genetic markers was used in the Micheletti et al. 2020 study examining the distribution of genetic ancestry in persons of African descent now living in the Western Hemisphere.

HOW IS ANCESTRY DETERMINED BY DNA? ARE GENETIC ANCESTRY AND RACE THE SAME THING?

Ancestry typically represents a generational narrative about one's relatives through their maternal, paternal, or both lines of descent. Sexual organisms such as ourselves display biparental inheritance. Our offspring are composed of diploid (2N) genomes whereby one copy (1N) of the genetic code originates from each parent. This means that in each generation, an individual has two ancestors. Going backwards through generations, this amounts to a maximum of 2^N ancestors, where N is the number of generations in the past (two parents, four grandparents, eight great-grandparents, and so on). Therefore, about half of your genome came from each parent, a quarter from each grandparent, and one eighth from each great-grandparent. The logic of this process means that all modern humans are related going back to the first humans who lived in sub-Saharan Africa 200,000 to 300,000 years ago.

Because our species began as hunter/gatherers, population growth was very low until the advent of agriculture and the Industrial Revolution. Estimates of world population in 1500 range from 425 to 540 million. Consider this in comparison with our current population level, which exceeds 7.8 billion people![5] Approximately ten human generations have passed since the year 1700. This means that you have had about $2^{10} = 1,024$ ancestors, assuming no inbreeding in your lineage. However, because everyone's genealogy contains marriages between cousins, your actual number of ancestors is considerably less than that.

Mitochondrial DNA (mtDNA) is a genetic marker that children inherit from their mother. The Y chromosome SNP haplotypes are passed from father to son. These tracing methods allow direct comparison of the lineages of sampled and referenced individuals. For mtDNA, the genetic markers are from the D-loop region of the molecule. This method has been used in forensics (more on this later in the chapter).

For example, mtDNA was used to identify the son of deposed French monarchs Marie Antoinette and Louis XVI. The boy supposedly died in prison at age ten. In 1845, he was given a royal burial, but some suspected that an imposter was buried instead. His desiccated heart was preserved by the royal family. In 2001, researchers utilized heart and hair follicle tissue from the bodies of Marie Antionette, two of her sisters, and the still living queen of Romania (a descendant of Marie Antionette) to determine that the body was really that of her son.[6]

Another historical debate was resolved using Y chromosome sequences. This was the long-standing claim concerning whether President Thomas Jefferson fathered the children of Sally Hemings, one of his slaves. Annette Gordan Reed recounted this saga in 1998.[7] A genomic study showed that Jefferson's Y chromosome matched Sally's last son, Eston Hemings, but did not match the Y sequence of Thomas Woodson, Sally's first son. This result is interesting in that Sally became pregnant with Thomas while she was Jefferson's servant in France. Given that France had outlawed slavery, Sally was a free woman in that country and could not be compelled to return to slavery in Virginia. Gordan Reed recounts that to get Sally to return to America, Jefferson agreed to free all of her children. This indicates that Jefferson must have thought that he was the father of this child.

The DNA results show only that someone carrying a "Jefferson" Y chromosome fathered Sally Hemings's last son, but it could have been another male in the Jefferson family.[8] This case also underscores the difference between socially defined race and genetic ancestry. Hemings was a half-sister of Thomas Jefferson's first wife. She was fathered by Jefferson's father-in-law and born to one of the enslaved women on his plantation. This woman also had at least half European ancestry, so Sally had three quarters European and one quarter African ancestry; yet by Virginia law she was still a slave. All of her children were fathered by men of European descent and thus would have been seven-eighths European and one-eighth African. Her daughter, Harriet, was also most likely fathered by Thomas Jefferson.

Harriet was eventually allowed to leave Monticello in 1822 at age twenty-one and moved to Philadelphia. She married a man of European descent. Harriet must have passed as white, as no one knew her mother was enslaved on the Jefferson plantation. Her children were therefore fifteen-sixteenths European and one-sixteenth African. If anyone had known of the ancestry of Harriet and her children, their legal racial status would have differed by their state of residence. Some states would have declared them white, but in Louisiana (with its one thirty-second rule), they would have been declared Black.

Finally, is modern genomic technology, which enables us to learn about an individual's ancestry, just another way of classifying humans into biological races? How can a DNA ancestry company say what percentage of your DNA is from each continent? Isn't that race? No, it is not. That is geographic variation divided into continents. It is not biological race.

The method used by DNA ancestry companies takes advantage of the distribution of genetic variations within and between groups and continents, as well as whether human groups can be considered unique lineages (due to lack of gene flow). The ability to look at millions of SNPs has not changed our conclusions that humans do not divide into biological races. *If anything, whole-genome studies confirm that modern humans do not contain biological races.* For example, sequencing studies have shown that the Khoisan hunter-gatherers are different from sub-Saharan Africans to a greater extent than the latter group is from Europeans and East Africans, and the Maasai people are more similar to Europeans than they are to the Khoisan.[9] Within Africa, the Sandawe are closer to Europeans than they are to the Hadza.[10] This shows that genomic data disagree with older physical race classification schemes, as several groups with African physical features (Khoisan, Sandawe, Hadza, and East Africans in general) do not group together by ancestry.

ARE SOME ANCESTRY-TESTING SERVICES BETTER THAN OTHERS?

This is hard to answer, because the genetic markers and algorithms used by these companies are proprietary. This means that not all of the DNA panels that these companies use to infer ancestry are open to scientific peer review. Therefore, we can only describe attributes of testing methods that would make them better estimates of "real" ancestry and in which populations ancestry testing should work best. Clearly, the more genetic markers used, the more repeatable the estimates of ancestry will be. (For example, in the studies cited earlier, 23andMe used more than two million markers.) The markers should also be well dispersed across the twenty-three pairs of human chromosomes. In this way, the SNPs mark more segments of each chromosome that are inherited together.

Also, it is important to know something about the populations being tested and their demographic history. In theory, the methods should work best for populations that have been relatively stable in size (no recent major reductions) and that have experienced little migration/emigration recently (which would introduce new markers or reduce the frequency of older markers). Unfortunately, human groups tend to change in size and to move and mix. So, the assumption of stability on which ancestry testing is based are shaky in some groups.

With these caveats in mind, certain socially defined groups are more likely to get more accurate DNA ancestry test results than others. Individuals whose ancestry is from large European groups (such as English, French, German, and Italian) and East Asian (Han Chinese, Japanese) that have seen little population reduction or gene flow should get results that are more accurate than persons with substantial western African ancestry. (The reason is that west African societies were disrupted by wars, and large numbers of individuals were removed by the transatlantic slave trade.)

This effect would differ by the specific western and central African group in question. During the transatlantic slave trade, growth in the total population of this region stagnated.[11] In other words, the birth rate equaled the death rate, for an intrinsic rate of increase (symbolized as r) that would have been zero or negative. However, the births and deaths in this region were not equally distributed, so some groups might have seen serious reduction of their numbers, with the resulting allele frequencies determined by genetic drift (random chance). Indeed, this is highly likely to have been true for the Amerindian populations of North America and the Caribbean (whose populations were drastically reduced by colonialism, war, and epidemics; see chapter 5). Ironically, African Americans or American Indians taking DNA ancestry tests are likely to get a more accurate assessment of their European ancestry that their western or central African or Amerindian ancestry.

Data on the within-company reliability of DNA ancestry testing indicate that applying the same set of genetic markers to an individual almost always produces similar results. That is because the same algorithm is used on the same data. A recent study examined the reliability of DNA-testing results of identical twins who were tested by the same company. Identical twins have identical genomes (although they might differ by small amounts due to somatic mutations accrued over their lifetime). The study examined results from 23andMe, Ancestry, and MyHeritage and showed that within-company tests had from 94.2 to 99.2 percent agreement between the identical twins.

However, when the twins were tested by different companies, the agreement among the estimates resulting from the different companies fell to only 52.7–84.1 percent.[12] This decline in replicability is due to the use of different data sets and algorithms. In addition, the three companies characterize ethnicity differently. For example, to describe ancestry from the British Isles: Ancestry used Ireland/Scotland and England/Wales as categories; MyHeritage used Irish, Scottish, Welsh, and English; and 23andMe used British and Irish. For West African ancestry, Ancestry had no panel, MyHeritage used Nigerian, and 23andMe used Nigerian, Coastal West African, Senegambian, and Guinean categories. In the latter case, the 23andMe categories are clearly more accurate than those of MyHeritage and Ancestry for western African ancestry.

Finally, a counterintuitive result of genetic ancestry testing is how it has undermined claims of racial identity by individuals whose racial identity was central to their worldview. Very early in the history of these DTC companies, Afrocentric individuals received results showing that their DNA indicated substantial European ancestry. We know that the flow of genetic information between Africans and Europeans in the Western Hemisphere was mainly unidirectional due to laws that defined racial status. One study found that the European genetic contributions were highest in African Americans for the Y chromosome (28.46 percent), followed by the autosomes (19.99 percent), then the X chromosome (12.11 percent) and mtDNA (8.51 percent).[13] This means that European males were far more likely to impregnate women of African descent (such as in the story of Thomas Jefferson and Sally Hemings) than vice versa. A 2015 23andMe study of socially defined European Americans found only 0.19 percent genome-wide recent African ancestry.[14]

The ubiquitous TV ads for genetic ancestry testing all sell these tests as somehow revealing some true (genetic) self. One of the most historically inaccurate TV ads was run by Ancestry.com in early 2019.[15] The spot revolves around a fictionalized man of European descent imploring a woman of African descent to escape with him from slavery. The reason he wants the woman to go with him is because she is carrying his child. Although it is entirely possible that some men of European descent had consensual love affairs with women of African descent during the time of chattel slavery, historical records clearly demonstrate that the vast majority of these pregnancies resulted from the rape of enslaved women.[16] After a massive public uproar, the advertisement was removed on April 15, 2019.

The fact that some persons of African descent managed to pass as white played out in the story of Davis Knight, the grandson of Newton Knight. Newton, played by Matthew McConaughey, was the main character in the 2016 film, *Free State of Jones*. Mississippi charged Davis with miscegenation due to his alleged one-eighth African ancestry when he married a white woman, Junie Lee Spradley, in 1946. Newton's second wife was reputed to be a Negro, and the state claimed that Davis was descended from this woman (Rachel Knight), not Newton's first wife. Davis challenged both the state miscegenation law and his racial status. He and his wife eventually left the state after Mississippi dropped the charges against him in return for his agreement to stop challenging the miscegenation law.[17]

Probably the most ironic case of DTC testing resulted when the DTC ancestry results of Craig Cobb, a white supremacist, were revealed on television. They showed that he had 14 percent African ancestry, an amount consistent with having one great-grandparent with 100 percent African ancestry. Of course, Cobb denied the validity of the results and eventually recast them in an attempt to rescue his white identity.[18]

HOW IS ANCIENT DNA (DNA OBTAINED FROM ARCHAEOLOGICAL AND FOSSIL MATERIALS) CHANGING OUR UNDERSTANDING OF HUMAN EVOLUTION?

Recent developments in technology are now allowing scientists to extract and analyze DNA, including ancient DNA (aDNA), from humans and formerly living organisms. The genetic material has now been extracted from relatively recently deceased individuals, such as historical figures, as well as from fossils that are tens of thousands of years old. Once an organism dies, its DNA begins to break down. For this reason, the fossilization conditions that allow adequate amounts of genetic material to be preserved are rare. In addition, there is frequent contamination of the sample with DNA that originated from outside the sample. A bone lying in the soil accumulates DNA from various microbes as well as from any individuals who might have handled the fossil in the recovery process.[19]

However, in the last decade, researchers have substantially improved both experimental and computational methods that correct for the effect of contamination on these ancient samples. The process includes extracting the sample's DNA under strict clean-room conditions, using UV radiation to degrade microbial DNA, bleaching all surfaces where the extraction takes place, and using filtered air systems. In addition, computational methods are used that allow researchers to identify the degradation associated with the ends of ancient DNA that is not found in modern DNA.

Over the last few decades, researchers have recognized that we cannot understand past human migration by studying only the genomes of the people who are alive today.[20] People move! The ancient DNA revolution has recently allowed geneticists and anthropologists to test models of ancient human migration in ways not possible a few decades ago. The new data show that human migration in the ancient world was complex, just as it is in the modern world.[21] Some researchers suggest that modern human populations are the result of the admixture (interbreeding) of highly differentiated existing populations. They further claim that these population admixture events occurred repeatedly in the evolutionary history of our species.

This view of ancient human migration is more complex than the serial founder effect model. In the latter model, small populations leave large populations to found new populations, which are dispersed from the ancestral population (see fig. 9.1). When modern humans began to migrate from Africa, our species' population size was estimated to be between fifty and one hundred thousand individuals.[22] This migration was possible because of changing weather patterns near the end of the Pleistocene era. Glacial ice sheets began to recede, allowing subgroups of modern humans to exit Africa. For example, one can imagine the populating of the Australian continent as beginning in East Africa around 50,000–60,000 years ago and progressing through the founding of subsequent populations in the Arabian peninsula, Indian subcontinent, Indochina, the islands of New Guinea, and, finally, to the Australian mainland. In the original model of human expansion, hybridization with other human species was not thought to play a major role.

On the other hand, in the more complex admixture model of the populating of the Australian continent by modern humans, hybridization with other human species plays a prominent role. Ancient DNA reveals that the individuals who founded the ancient Australian populations had genomic elements from anatomically modern humans: Neanderthals, Siberian Denisovans, and a group dubbed Australo-Denisovans.[23] These were Denisovan humans who lived southeast of Huxley's line, a demarcation between animals undergoing adaptation in mainland Asia and those undergoing adaptation

A₁, A₂, A₃, A₄, A₅, A₆, A₇, A₈

Original large population contains
8 alleles at a given locus.

A₁, A₂, A₃, A₄, A₆, A₇, A₈

First founder population loses
A₅ allele by chance alone.

A₁, A₇, A₈, A₄

Second founder population loses
A₂, A₃, and A₆ by chance alone.

A₁, A₄

Third founder population loses
A₇ and A₈ by chance alone.

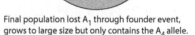

A₄

Final population lost A₁ through founder event,
grows to large size but only contains the A₄ allele.

FIGURE 9.1. Potential population genetic effects of the serial founder effect. The initial large population contains eight alleles. Alleles are lost as each smaller population migrates to new territory, ending with only one allele (A₄) remaining at the end of the migratory journey. As a result, the genetic diversity of each serial founded population declines with distance from the original large population. (The founder population would have thirty-six possible genotypes, whereas the final population would have only one genotype, A₄A₄.)

in the transitional ecological zones from Asia to Australia. These different hominid species were separated by hundreds of thousands of years of evolution, a much greater time span than the separation between San hunter-gatherers and the rest of modern humans (about 70,000 years).

Thus, during the migration of our species around the world about 100,000 years ago, different groups encountered and mated with other groups, archaic human species, as they traveled. The decrease in the percentage of archaic DNA that originated from these species in modern humans results from the fact that these other species went extinct. Therefore, mating with them ceased. By genetic recombination, each generation would have the amount of archaic DNA reduced by 50 percent. Thus, in the genome of modern East Asians, Denisovan DNA is about 0.2 percent and in South Asians, 0.3–0.6 percent.[24]

Some of the genes that were acquired by hybridization with other species were beneficial. There is evidence that in Tibetans, the genomic variant associated with high-altitude adaptation is more closely aligned with ancient Denisovan DNA than it is with the modern human sequence of that gene.[25] Movement of genes between species resulting from hybridization is well known. Generally, in the case of "true" species, hybrid individuals have lower evolutionary fitness compared with their parental species. An extreme case of this is the mule, the offspring of a horse and a donkey. Most mules are sterile.

However, in many closely related species, the hybrids are only slightly less viable than the parental species, as in the case of the fruit fly species *Drosophila melanogaster* and *Drosophila simulans*. If a *D. melanogaster/simulans* hybrid mates with either a *D. melanogaster* or a *D. simulans* individual, genes that originated from one of the species can move into the other. If the new gene is beneficial compared with the ancestral one, natural selection will increase its frequency within the species that acquired the new gene. An example of this has been shown recently with killifish. Adaptive toxicant resistance rapidly evolved in Gulf killifish (*Fundulus grandis*) that live in polluted habitats. There is a clear relationship between the amount of resistance in the fish populations and the pollution level they experience. Two loci with the strongest signatures of recent selection encode genes that provide resistance to the specific toxin the fish experience, and the resistant alleles moved between the species recently (eighteen to thirty-four generations ago) from the Atlantic killifish (*F. heteroclitus*).[26]

Adaptive introgression would account for the high frequencies of genes that originated from Denisovans or Neanderthals in some modern human populations. Of course, it must also be understood that parallel or convergent evolution could have easily produced the high-altitude genetic adaptation displayed in Tibetans. In that case, it is likely that the Tibetan variant originated in Denisovans.

This was not true of the Andean and Ethiopian high-altitude variants of this gene, which originated in modern humans. Indeed, to give you a sense of the power of convergent evolution, the cold temperature adaptations observed in ancient Neanderthals are also found in mastodons![27] Mastodons are extinct members of the same order as modern elephants (*Proboscidea*). Both species originated in tropical conditions, and some of their subpopulations migrated to colder climates. As the temperature regulation mechanisms of mammals are deeply conserved, both Neanderthals and mastodons would have begun their adaptation to colder climates utilizing similar genetic mechanisms.

By the beginning of the twentieth century, naturalists had come to understand that all species were composed of populations (groups of organisms inhabiting specific space and time) that exhibited a continuity of interbreeding across their range. Modern humans are somewhat unique among land mammals, in that our range extends over most of the planet. The only place that does not feature indigenous human beings is Antarctica.

There is general agreement that our species originated in sub-Saharan Africa. Therefore, what is at issue now is our exact understanding of how modern humans expanded across the planet. The current consensus is that modern humans (*Homo sapiens sapiens*) originated in eastern, central, or southern Africa around 300,000 years ago. The first fossil evidence of modern humans found outside Africa dates from 100,000 years ago in the Middle East and 80,000 years ago in southern China.[28] There are several ideas about how and when modern humans left sub-Saharan Africa, with estimates ranging from 50,000 to 100,000 years ago. Some debate about whether there were multiple waves of migration, as well as by which route the migration occurred (northern or southern).[29] It is also clear that during these migrations, modern humans hybridized with various species of archaic humans in Africa, Europe, and Asia.[30]

Modern humans arrived in Europe around 43,000 years ago, but most likely these first Europeans did not contribute much to the people who currently live in Europe, whose ancestors arrived around 7,000–10,000 years ago.[31] The first Europeans were primarily hunter-gatherers, and these populations were replaced by humans who had developed farming.[32] A third wave of migration into Europe occurred about 4,500 years ago (Early Bronze Age). These migrants were descendants of hunter-gatherers who originated in what today is modern Russia and the Caucasus mountains.[33]

Asia was populated by at least two early waves of migration: first by ancestors of Australians and the Papuans, followed by other ancestors of East Asians with modern populations resulting from admixture between all the individuals and groups that were parts of these two waves of migration.[34] However, many details about migration into Asia are yet to be resolved. These details may emerge with further studies of aDNA. At present, the archaeological data place modern humans in Oceania 47,500–55,000 years ago. The extensive genomic study of Papuan and aboriginal Australians suggests that genetic divergence between these groups might have been driven by climate changes and that the aboriginal Australians were living in a relatively high level of isolation until modern times.[35]

Finally, the earliest arrival of modern humans in the Western Hemisphere is accepted to be about 14,000–15,000 years ago. The widespread occupation of that region is associated with the Clovis culture (12,600–13,000 years ago). It is likely that the genomic divergence of Siberians and the Amerindian populations began around 23,000 years ago.[36] This suggests that Amerindians moved into the Western Hemisphere earlier than the dating of archaeological remains indicates; however, by what means and route this occurred is still the source of considerable controversy. The earliest archaeological sites could be buried under ice or in the Pacific Ocean (because the ancient shoreline has changed) or simply not found and accepted by the archaeological community.

The migratory paths of humans around the world led to divergences in their gene frequencies, primarily driven by genetic drift.[37] In addition to drift, great distances between populations reduced the amount of gene flow. Also, as humans migrated away from sub-Saharan Africa, new environmental conditions, including extreme cold, altered exposure to sunlight intensity, and new pathogens were experienced. This also occurred as humans moved around sub-Saharan Africa.

As noted in chapter 2, Africa is the second-largest continent, covering more latitude than all other landmasses (about 37° N to 35° S) with highly diverse vegetation zones. However, at the end of the Pleistocene, the Sahara was reduced in size.[38] As we previously noted, human populations that migrated within Africa faced new environmental conditions and habitat diversity, and this partly explains why sub-Saharan Africans display greater genetic diversity than all other populations. Also, throughout human migration, cultural evolution introduced new methods of hunting, along with plant and animal domestication, creating changes in diet.[39] That process leads to the genetic structure within the human species that we see today. However, modern human populations are not just the result of ancient migrations. Events such as the expansions by the Bantu and Mongol Empires and the transatlantic slave trade have altered populations in the last thousand years or so. These events altered the genetic structure within our species, as they increased the gene flow between populations located in different regions of the world.

In summary, ancient DNA has provided us with a new perspective on how human populations expanded and merged around the world. On the global scale, the best way to understand human genetic variation is to recognize that it was driven by genetic drift resulting from serial founder events. Interspersed with genetic drift was adaptation to new local conditions. Because ancient migration was not unidirectional, gene flow among human populations maintained genetic uniformity, such that our species never formed biological races. Finally, the coming of technological and cultural innovations in the modern age accelerated gene flow among global regions, further reducing the possibility that biological races would ever form in modern humans.

HOW CAN MORPHOLOGY AND DNA BE USED IN FORENSICS?

One of the main goals of forensic science is to identify individuals who are either victims of crime or perpetrators. Typical FBI forms for filing victim reports include questions about estimated age, sex, height, and "race" of the perpetrator. In cases in which only bones and teeth are present, a forensic anthropologist is typically called in to make all of these determinations. Standard methods are available to estimate height (based on long-bone lengths), sex (based on features such as the size and shape of the pelvis), and age (determined from multiple signs of aging in teeth and bones). But what about race?

Forensic anthropologists have made a number of suggestions that race can be determined from skeletal remains. These include the size and shape of the cranium, harking back to old studies of cranial shape and intelligence.

What we call racial anthropology is an anthropology that maintains the fiction of biological races and is obsessed with determining differences among biological races. Many forensic anthropologists are still heavily into racial anthropology.[40] They might argue that this is because forms from the FBI ask them to fill in a victim's race. The problem is, of course, that race is a biological fiction. And it is reinforced by the fact that as much as forensic anthropologists attempt to reliably determine race from bones and teeth, they continue to fail to do so.

A good example of this failure revolves around the use of discriminant functions of skull shape as a way to distinguish sex and race. The gold standard method was developed by Giles and Elliot and published in 1962.[41] These two, then anthropology graduate students at Harvard, determined that a set of measurements could distinguish males from females. Then another set of measurements could distinguish among three races: Black, white, and American Indian. They were able to do so with almost 90 percent accurate "racial classification" in their original study. So far so good.

The problem comes in efforts to apply their method to different locations and samples. Alan studied this problem and located four restudies. He found that on average, the American Indians were correctly classified in the four restudies about one third of the time and misclassified as white or Black the remaining two thirds of the time.[42] Given that random classification would return about the same results, the method is no better than chance when it comes to determining "race." It's not even good enough for government work.

Why were retest results so bad? These data point to a couple of problems of racial anthropology. First, these are efforts to biologize a social category of race. They assume that there is a good fit between the two, but there is not. Second, the studies assume that race is universal and the same in all times and places. But that's not true. Indeed, a Black or white person in St. Louis and Cleveland in the twentieth century is different socially and genetically from a Black and white individual elsewhere at other times. Color lines change.

More recently, DNA evidence has begun to supplant phenotypes such as skull shape in efforts to determine the identities of victims of crime and especially the perpetrators. Previously, efforts were made to study any sort of information that might have been left behind by a possible perpetrator; for example, footprints or fingerprints. DNA has been shown to be more precise in making these determinations. For example, if a 100 base pair segment of noncoding DNA is used, the possible number of sequences is 4^{100}, or about 1.6×10^{60} possibilities! This is greater than the number of stars in the known universe. To identify some of the victims of the World Trade Center bombing, Myriad Genetics utilized thirteen short tandem repeat (STR) loci. Short tandem repeats are segments of DNA (2–7) base pairs long.[43] The chance that any two individuals share the same thirteen STR genotypes is 1 in 250 trillion (2.5×10^{14}). This means that no two people on Earth (other than identical twins) can share the same set of thirteen STR markers.

Now, with millions of individuals in DNA databases, we have approached a time when DNA left at a crime scene, perhaps from a single hair, can be sequenced and matched. The matching need not be to the individual; it could even be to a potential cousin or more distant relative. This is precisely how the Golden State Killer was identified.[44] Conversely, the lack of a match between the DNA of someone previously incarcerated with the genetic evidence found at the crime scene is leading to overturning previous convictions, most often of Black men.[45]

However, there is a dark side to DNA databases. Specifically, many states retain DNA profiles of people who were accused of a crime but never convicted.[46] In the United States, you are presumed innocent until proven guilty. Unfortunately, disproportionate numbers of racially subordinated people

(African Americans, Latinos) are arrested. As a result, the genetic information of these innocent individuals and, by extension, their families remains in DNA databases.

In 2008, the European Court of Human Rights ruled that the indefinite retention of the DNA of convicted and innocent persons was disproportionate. However, to get around this ruling, in 2012, England and Wales passed the Protection of Freedom Act, which allows the indefinite retention of the DNA of convicted persons and the temporal retention of DNA of first-time convicted minors and innocent individuals in the country's database.[47]

Recent events in China are even more disturbing. There the government is collecting DNA from men and boys across the country to build a database targeting ethnic minorities and other groups. This database will allow the tracking of individuals (and their families) from any of their biological material. This is being accomplished with the help of American biotechnology company ThermoFisher.[48] What should be clear is that this recent use of DNA in forensics has little to do with race. However, DNA databases are currently being deployed in racist ways. Their use is consistent with the existing patterns of social dominance in various nations, resulting in intensifying the disparity in social justice in these societies.

CONCLUSIONS

In this chapter, we have tried to answer questions about what ancestry is and have presented three types: cultural, geographic, and genetic. Just as social race does not always map onto biological ideas about human variation, so, too, cultural, geographic, and genetic ancestries do not always map onto one another.

Recreational genomics, and specifically genetic ancestry testing, is now a large and expanding industry. We explained how genetic ancestry might be determined and especially why being able to do so has nothing to do with biological race. Rather, these tests take advantage of the fact that there is geographic and population-level variation in genomes. The key problem is that the ancestry companies often give a false sense of precision, and the results are interpreted in racial terms. This shouldn't be a surprise, given that biological race remains such a dominant worldview. That's why it is important to be clear that ancestry and biological race are very different constructs.

Studies of past individuals, directly through bones and teeth and their archeological remains and now through their ancient DNA (aDNA), provide a clearer view of human history. Perhaps the biggest lesson we are learning is that humans always moved and merged. Thus, what might be called English at one time might look very different physically from English today or at any other time. Moreover, there is no reason to continue to think that social definitions of race neatly map onto the mythic idea of biological races. Finally, as W. E. B. DuBois said well over a century ago, the problem of the twentieth century is the problem of the color line. Indeed, the color line is still shifting and continues to be a problem well into the twenty-first century.

Chapter Ten

RACE NAMES AND "RACE MIXING"

Some years ago, psychologist Jefferson Fish brought his daughter and her Brazilian boyfriend to his class at St. John's University in New York. Professor Fish had often accompanied his wife, anthropologist Dolores Newton, to Brazil, where she did her fieldwork. Fish noticed that his wife, an African American, seemed to become white when she got off the plane and entered Brazilian society. In his cross-cultural psychology class, Fish, the son of Eastern European Jewish immigrants, asked his daughter and her Brazilian friend point blank if they were Black. His daughter said without hesitation that she was Black. Her Brazilian friend, who is considerably darker-skinned than her, said no.[1] How can that be so?

This chapter examines how we have talked—and how we should be talking—about socially defined race, ancestry, and identity. We discuss the use of race as a census question in the United States, other forms of racial classifications, ideas about "racial" purity and whiteness, and confusing racial categories. We examine how and why the U.S. government collects data (head counts) by race and ethnicity and the differences and similarities between official data and everyday, street-level classifications. We discuss changes in racial classifications over time and variation by countries, as in the comparison between the United States and Brazil.

In race classifications, we find both an old, color-coded system that harkens back to Linnaeus and the early classifiers and, at the same time, one that changes over time and in different locations. Who is classified as white changes, over time in Europe and the United States, and over space, as when Professor Newton boards a plane from New York to São Paulo. The color lines shift, but they are still there and remain heavily guarded. Race is both stable and a chameleon that conveniently changes to fit the local political and social power structure.

WHY DOES THE UNITED STATES COLLECT CENSUS DATA BY RACE?

First, we collect data on the social race of those living in the United States because in 1787, Congress mandated that we do so. But that, of course, is just the first, superficial pass at an answer. Why we continue to abide by the law has changed through time. In 1790, the year of the first census, we counted the number of free white men over sixteen years of age in order to establish taxes and representation in the new states. In addition to counting the white men, it was thought, why not count slaves and women? We did that. Infamously, in one of the oddest of compromises in U.S. history, enslaved Africans were counted as three-fifths of a person. That compromise gave the slaveholding states greater representation in Congress than would have been allowed by their white male population alone. As such, this was one of the original hypocrisies of American social life. The war for independence had been fought based on taxation without representation. Yet, enslaved Africans were counted to bolster the representation of their owners, not for the slaves' benefit.

Now, more than two hundred thirty years later, we continue to count all citizens by race, not only in the decennial census but for most federal programs including housing, employment, health and health care, and education. We do so in order, at least in theory, to provide equal representation and track inequalities by race as well as ethnicity. Oddly, in federal data, Latinx individuals are considered to be an ethnicity. For example, collecting health data by race allows us to track inequalities in infant mortality and life expectancy as well as educational attainment and to track labor practices and housing loans.

There is a potential contradiction at the heart of collecting data by race: we thereby affirm that race must be a useful and real category. And socially defined race is real. However, many continue to think that biological races are real. However, as we have shown in prior chapters, humans do not have biological races, but race is real because people have made it real. Moreover, the consequence of being asked one's race suggests that race categories are fixed and stable when, in fact, they are anything but.

The official census categories show just how unstable race categories are. In 1790, the categories were Indian, slave, and white. In 1810, the category of free colored persons was added. Later, categories such as mulatto (half Black and half white, 1850), quadroon (one-quarter Black and three-quarters white), and octoroon (one-eighth black and seven-eighths white) were added in 1890. Asians—specifically Chinese—were added in 1870 and Japanese in 1890. Mexican, even though many had been living in the United States since 1848, was added in 1930 and removed in 1940.

In 1970, a question about Hispanic ethnicity was added. Note the term "ethnicity." As a group of individuals who generally result from the intermixing of individuals from the Iberian Peninsula, indigenous Americans, and enslaved Africans, it must have seemed wrong to call Hispanics a race. If so, what race are they? The two largest groups of Hispanics in the United States are Mexican Americans (who are more American Indian than European, with little African descent) and Puerto Ricans (who have less American Indian, about the same European, but more African descent than Mexican Americans). The percentage of these three ancestry groups in Caribbean Hispanics can differ widely.

Of course, by that reasoning, it is just as wrong to call Blacks or African Americans a race, as this group also has African (about 83 percent), European (16 percent), and indigenous American ancestry (1 percent; see chapter 9). It seems the primary reason for identifying "Hispanic" as an ethnicity is that many of these individuals come from cultures in which Spanish was the primary language. So, the U.S. census now includes the option to select Hispanic ethnicity and a race of one's choosing.

Slowly, the race categories expanded further (fig. 10.1) and now include multiracial options. Sometimes names change to keep up with changes in common usage. However, it is clear that the census categories have always been driven by the political needs of the government. They have nothing to do with the biological ancestry of the people being classified.[2] They are also not backed by science and anthropology. In fact, in the Office of Management and Budget Directive 15, the section in which the categories are laid out and defined specifically states that they are "neither scientific nor anthropological"![3]

Finally, the United States is among a minority of countries that are obsessed with recordkeeping by race. For example, in the United Kingdom, data on class and occupation are much more available. One can better track health changes there by class rather than race, for instance; the opposite is

Race categories	1790	1800	1810	1820	1830	1840	1850	1860	1870	1880	1890	1900	1910	1920	1930	1940	1950	1960	1970	1980	1990	2000	2010
Indian	■	■	■	■	■	■	■	■	■	■	■	■	■	■	■	■							
Slave	■	■	■	■	■	■																	
White	■	■	■	■	■	■	■	■	■	■	■	■	■	■	■	■	■	■	■	■	■	■	■
Free colored person			■	■	■																		
Black							■	■	■	■	■	■	■	■	■	■	■	■	■	■	■	■	■
Mulatto							■	■	■	■	■		■	■									
Chinese								■	■	■	■	■	■	■	■	■	■	■	■	■	■	■	■
Japanese									■	■	■	■	■	■	■	■	■	■	■	■	■	■	■
Octoroon											■												
Quadroon											■												
Other													■		■	■	■	■	■	■	■	■	■
Mexican															■								
Negro												■	■		■	■	■	■	■	■	■	■	■
American Indian															■	■	■	■	■	■	■	■	■
Filipino															■	■		■	■	■	■	■	■
Hawaiian																		■	■	■	■	■	■
Korean															■	■			■	■	■	■	■
Aleut																				■	■		
Asian Indian																				■	■	■	■
Eskimo																				■	■		
Guamanian																				■	■	■	■
Hispanic*																				■	■	■	■
Samoan																				■	■	■	■
Vietnamese																				■	■	■	■
Other Asian or Pacific Islander																				■			
African American																						■	■
Alaska native																						■	■
Chamorro																						■	■
Latino*																						■	■
Other Asian																						■	■
Other Pacific islander																						■	■

FIGURE 10.1. Official U.S. census categories by decade. *Source:* A. H. Goodman, Y. Moses, and J. Jones, *Race: Are We So Different?*, 2nd ed. (New York: Wiley, 2019).

true in America. This makes sense given the enduring dynamics of class discrimination in the UK and racial discrimination in the United States. However, it obscures the lived significance of social class in America and the lived significance of racism in the UK.

Still, although imperfect, the collecting of data by socially defined race is very important because it is the only way we can track racial inequalities and, we hope, progress toward equality.

ARE RACIAL CLASSIFICATIONS THE SAME IN DIFFERENT COUNTRIES?

Yes and no.

We first say yes because the classification of three to five main races that has come down from natural scientists such as Linnaeus and Blumenbach still dominates. These racial classifications or names sometimes correspond to continents, but not always. Some variation, sometimes small and sometimes a bit more significant, are found in almost every place we are aware of.

For example, race classifications in European countries are rather similar. Racial distinctions are almost always made between European, African, and Asian derived populations. In 2019, Joe gave a lecture in Poland and was informed that the public education system there still taught the three-race classification scheme as if it were state-of-the-art science. Sometimes Native Americans are lumped in with Asians, sometimes they are ignored, and sometimes they are added as a separate race. Native Americans are a separate race in the United States because their histories are very different from those of other Asians. Sometimes Native Australians, Pacific Islanders, and Middle Easterners are added as separate races.

Racial classifications tend to differ more outside the West, where there are different histories of colonization and slavery. Race might even be less important. For example, in India, a country that did not experience the importation of Africans or other non-Indians as laborers, the caste system functions a bit like a mix of race and class.[4] Not unlike race, one's caste is inherited and passed down through generations. Castes are also like classes, in that they are tied to occupations.

However, Western ideologies have spread throughout the world, and part of that ideology is the myth that biological races are real. Thus, we still almost always see the big three races—European, African, and Asian—referred to in almost every corner of the globe.[5] Classifications also take on a more local flavor based on the dynamics of different locations. For example, throughout Central and South America, physical signs that we hold to be signifiers of race, such as hair form and skin color, merge with signifiers of social class, such as education and economic status. Socioeconomic class tends to correlate with skin color (and admixture). Lighter-skinned individuals, the descendants of conquering Europeans, tend to be more educated and wealthier. This is also why Dolores Newton, the African American anthropologist mentioned earlier, becomes whiter when she travels to Brazil. This dynamic is largely true in Puerto Rico, where any European ancestry turns you white, and in Brazil, where there are several categories of race that would be classified as Black in the United States.[6] In addition, the amount of "Indian-ness" is more important in countries such as Mexico, Guatemala, Bolivia, and Peru, all of which have deep histories of admixture.

So, our racial classifications are similar throughout the world because of the dominance of Western racial classifications. And they also vary over time and place as chameleon-like adaptations to the needs of local power structures.

WHAT IS THE "ONE-DROP RULE"?

The one-drop rule is an informal classification traditionally followed in the United States whereby having a single drop of "Negro" blood makes one a Negro. It is an extreme version of *hypodescent*, a practice in many cultures by which children of mixed race are classified as members of the less dominant race. The one-drop rule is a guide to social classification. It is explainable if you believe that African ancestry is a pollutant. Interestingly, Negro blood is the most potent type and able to pollute a white body, whereas this is not as true for the blood of other races.

The one-drop rule is anachronistic. There is no such thing as Negro blood or blood of any so-called race. However, this as well as other rules of hypodescent have had great historical importance and continue to be observed in informal yet powerful ways.

The percentage of inheritance (or blood) from a Negro was often used to determine the legality of marriage across races. Protecting the color line or, more accurately, the white race, was paramount. Thus, at different times, it became important to be clear about the percentage of Negro blood in each person and to guard against a Black with only a drop of Negro blood passing as white (see chapter 9). A drop of Negro blood was defined differently by state. In Louisiana, it was one thirty-second, or just one great-great-grandparent. Other states allowed it to be one-quarter (one grandparent) or one-sixteenth blood (a great-grandparent).[7] The consequence of different legal thresholds for classification as white or nonwhite meant that a person could change their race simply by moving from one state to another. Note that we use the term "blood," not "genes" or "genetics." That is because at this time— the early 1800s through into the twentieth century—the nature of genetic inheritance had not been fully developed. Eugenicists became obsessed with protecting and maintaining the purity of the white race.

Many individuals have traced the origins of the Nuremberg Laws of 1933 and 1935, which classified individuals as Aryan or Jewish, to the U.S. one-drop rule. In Germany, individuals with partial Jewish heritage, generally one or two grandparents, were called *Mischlinge* (or mixed). One required a certificate of German ancestry to join the Nazi Party.[8] Interestingly, Nazi Germany's Nuremberg Laws allowed for some flexibility in determining

Jewish blood or parentage. Having one or even two Jewish grandparents did not automatically classify one as a Jew; thus, one could say that Jewish blood was less powerful than Negro blood.

The effectiveness of hypodescent statutes and cultural practices is illustrated by the fact that socially defined whites in the United States have on average just 0.19 percent recent African and 0.18 percent American Indian ancestry.[9] Of course, the opposite was far from true; due to rape and sexual coercion during chattel slavery, African Americans have on average 16 to 24 percent European ancestry (see chapter 9).

Today, the U.S. state and federal governments have overturned all legal references to the one-drop rule. All of the thirty-three states that had outlawed marriages across races, so-called miscegenation laws, overturned these laws mostly between the end of World War II and 1967, when the Supreme Court ruled them illegal.

Although the U.S. census and other legal documents no longer classify individuals by their percentage Negro blood, the one-drop rule had many lasting effects. Some individuals with parents from different ancestry groups can now call themselves multiracial. There is now a multiracial category on the census, and it is expanding. However, the ancestral origins of these multiracial individuals are still influenced by America's racial hierarchy. Individuals with a European American father and East Asian mother are far more common than those with a European American father and African American mother (despite Hollywood's and Madison Avenue's best intentions). But when push comes to gentle shove, most individuals with an African American parent or grandparent identify themselves, and are indexed by others, as Black. This is true of Barak Obama (whose father was Kenyan, not African American) and Kamala Harris (whose father is from Jamaica and mother is from India). Harris, affirming her Black identity, went to Howard University, a prestigious historically Black college.

WHAT DO I CALL PERSONS WITH AFRICAN, LATIN AMERICAN, AND ASIAN ANCESTRY?

We have repeatedly pointed out that humans don't have biological races. Thus, what we "call" people results from cultural, historical, and social forces. These forces are constantly moving, so group names within societies change. For example, Joe is "colored" on his 1955 birth certificate. This was the term in use across the country for children of African descent in the mid-1950s.

As a child, he identified as "colored" until 1968, when James Brown released *Say It Loud: I'm Black and I'm Proud*.[10] Listening to Brown was an awakening for Joe, a moment when he and many of his peers began to identify with the Black Power Movement.

Like many, later in his, life Joe shifted his identity to "Afro-American." This was spurred by his recognition of the significance of his African heritage. Today, Joe calls himself an "African American," again recognizing his African cultural and genetic roots and the significance of his family's American experience in forming his identity. The American experience also includes his European genetic ancestry through known ancestors; he estimates that he has about 18.75 percent European ancestry (see chapter 9).

We have previously discussed the confusing inadequacy of the term "Hispanic." Those who fall into that census category come from all over the Southwestern United States, Caribbean, and Central and South America. A historical fact that gives perspective to this discussion is that most people in this category have significant amounts of Amerindian ancestry.[11] Additionally, Spain was the first European power with colonies in the Western Hemisphere. Spaniards began their exploration of North America as early as 1513 with Ponce de Léon; Francisco Vázquez de Coronado explored the Southwest in 1541; and Pedro Menéndez de Avilés founded St. Augustine, Florida, in 1565.[12]

By comparison, Jamestown, Virginia, the first permanent English settlement, was not founded until 1607.[13] Furthermore, Mexico had established cities in the Southwest that were occupied by the beginning of the eighteenth century. Mexico outlawed chattel slavery in 1829. Mexico was conquered by the United States in the Mexican-American War of 1846–48, after which slaveholders in the United States moved into these teritories in order to expand slavery into the Southwest.[14]

Many Americans seem to forget that the residents of the island of Puerto Rico are U.S. citizens (Immigration and Nationality Act of 1952, § 302, 8 U.S.C. § 1402).[15] Similar to the Mexican southwest, Puerto Rico became part of the United States through an act of war against Spain; the 1898 Treaty of Paris ceded Guam, the Philippines, and Puerto Rico to America.[16] Paralleling the struggle for African American liberation, anticolonial and antiracist struggle among those in the census category of "Hispanic" has altered the way they think about themselves and therefore what they wish to be called.

Our general rule for addressing individuals in the Hispanic category is to ask them how they wish to be addressed. Some Mexican Americans prefer the term "Chicano," others prefer the generic terms "Latino, Latina, or Latinx." Still others wish to be referred to by their country of origin, such as Dominican or Cuban. The process of asking someone how they wish to ethnically and culturally self-identify demonstrates that you have an interest in their story. You might find out a great deal by simply asking this question.

Finally, the term "Asian" is possibly even more inadequate than "Hispanic" to describe persons with ancestry in this census group. The continent of Asia spans 17.21 million square miles, and depending on one's perspective, it includes the geographical regions including the Middle East, Eurasia, India, East Asia, and some islands of the Indian and Pacific Ocean. As we write, the population of Asia is around 4.5 billion people or about 60 percent of the world population. Thus, the label of "Asian" encompasses a huge variety of groups without specificity.

The 2020 U.S. census form lists the following options: "Chinese, Filipino, Asian Indian, Vietnamese, Korean, Japanese, Other Asian—Print, for example, Pakistani, Cambodian, Hmong, etc."[17] According to the U.S. Bureau of the Census, persons identifying as Asian numbered about 21 million (about 7 percent of the population).[18] Although Chinese and Japanese immigrants began to arrive in significant numbers in the nineteenth century, most Asian Americans entered the country after 1965. The largest groups are Chinese, Filipino, Indian, Vietnamese, Korean, and Japanese Americans (in that order). Our general rule applies here as well: the best practice is to ask individuals who fall within this category how they wish to be addressed. For example, Joe's wife simply refers to herself as "Korean."

In conclusion, the old race classifications are generally inadequate for capturing the lived experiences of individuals or how they self-identify. Classifications require more specificity and can change over time and by place. Finally, as a general rule, if you are unsure of how someone wishes to be identified, it is okay to ask.

WHAT WILL SOCIETY CALL MY CHILDREN?

American society remains mired in institutional racism. Underlying societal racism is the dominance of biological essentialist ideas held by the majority of Americans. A core notion of biological essentialism is the belief in the reality of biological races and the notion of racial purity. Even DNA ancestry testing, with results showing that almost everyone is mixed, fails to improve the general level of genetic essentialism among Americans. A recent study of U.S.

whites showed that for people with high levels of understanding of genetics, direct-to-consumer (DTC) ancestry testing did not reduce their level of genetic essentialism. Conversely, for people with low levels of prior genetic knowledge, such testing actually increased their level of genetic essentialism.[19]

Surprisingly, despite the centrality of race in American social life, the twentieth century saw the loss of legislation that defined races. Louisiana was the last state with a statute that defined "Black." That law played a major role in the case of Susie Guillory Phipps, who, in 1983, sued to have her racial status changed to white and to overturn the one thirty-second African descent rule.[20]

Throughout our history, the courts have used various racial notions to decide cases in which racial identity was a central factor. These include what we can call (a) status race, whereby the central premise is of Black inferiority and allows private individuals to discriminate on the basis of this belief; (b) formal race, holding that race is different based on appearance; (c) historical race, whereby race depends on an individual's or group's historical relationship to oppression and unequal power in society; and (d) cultural race, prioritizing a group's culture, community, and consciousness.[21] The lack of consistency allowed courts and judges to make rules that fit their biases and need to retain power and protect the white race.

It is probably even more shocking that cases can still be decided on the "status race" criterion. For example, in 1992, a white woman was awarded full worker's compensation disability for her phobia of Black men. She claimed to suffer from post-traumatic stress disorder because she had been mugged by a person she believed was Black.[22] Of course, one can only wonder why her PTSD was not just about men, as crimes by men against women are more common than those by Blacks against whites.

Title VII of the Civil Rights Act of 1964 directs the federal government to collect data on race and ethnicity. The current categories, their definitions, and the means of collecting these data are determined by the Bureau of the Census and the Office of Management and Budget (Directive 15). The OMB categories are as follows:

- "*White*: A person having origins in any of the original peoples of Europe, the Middle East, or North Africa.
- *American Indian or Alaska Native*: A person having origins in any of the original peoples of North and South America (including Central America) and who maintains tribal affiliation or community attachment.

- *Black or African American*: A person having origins in any of the Black racial groups of Africa.
- *Asian*: A person having origins in any of the original peoples of the Far East, Southeast Asia, or the Indian subcontinent including, for example, Cambodia, China, India, Japan, Korea, Malaysia, Pakistan, the Philippine Islands, Thailand, and Vietnam.
- *Native Hawaiian or Pacific Islander*: A person having origins in any of the original peoples of Hawaii, Guam, Samoa, or other Pacific Islands."

State agencies often have to decipher the self-reports that they receive for federal reporting requirements. For example, in 2004, the North Carolina Department of Public Health collected data on self-reported race at birth. From 118,000 live births, the mothers (or their representatives) reported 600 versions of racial categories for their children. As the federal government does not use 600 categories, the state workers collapsed these reports into the ten standard categories. This was not accomplished in an unbiased manner. Two-thirds of the Hispanic mothers reported the race of their baby as "other." As a result, the workers reassigned these children to the white racial category.[23]

Robert Hahn studied linked birth and death certificates. He found that over 40 percent of infants who died in the first year and were listed as Native American on their birth certificates had their race changed to white on their death certificates. He attributes this to the fact that although the mother was present at birth, the coroners might have filled out the death certificate without knowing the ethnicity of the mother or father.[24] What especially worries us about this example is that it directly implies a serious undercounting of Native American infant mortality.

With the decline of laws forbidding interracial marriage, there has been an increase in the number of marriages between people of different socially defined racial groups. This increase is nowhere near as large as one might expect from the recent glut of television commercials featuring such couples and their children.[25] A recent study of 543 advertisements found that the percentage of these couples in ads was much greater than the percentage of such marriages in U.S. society. In addition, nonwhite male/white female couples were underrepresented. No commercial with this dyad was found on the Disney Channel.

In 2013, about 3.1 percent of Americans claimed "interracial" ancestry. Within this group, the most common dyads were Black/white (> 2.4 million), followed by Asian/white (> 1.8 million), and Native American/

white (> 1.7 million). However, among these individuals, 61 percent did not identify as "multiracial."[26] In response to the question about how they identified themselves, most responded that they identified with the "race" that they most physically resemble. This is easy to explain, as the racial essentialism in which these individuals are immersed tend to generate that result.

Alan is friends with a couple, a white woman and a Black man. They refer to their daughter as Black (she has dark skin) and their son as Caucasian (he has much lighter skin). Joe has seen this in his own sons. One does not identify at all as African American. His appearance is closer to that of his mother, but he also does not identify as Korean. The other son identifies as African American (his skin is darker and his hair is curlier than his brother's). He feels the racism directed at his Blackness more acutely than his brother does. These responses differ by the dyad of the parents, as East Asians have social and cultural experiences different from those of African Americans. One study found that Asian Americans are actually more likely to classify multiracial (white/Asian) individuals as an out-group and not Asian. This, of course, would lead such individuals to have a greater tendency to identify as white, but that is counterbalanced by the lack of acceptance by whites of any multiracial individuals as white.[27]

So, the answer to the question about what people will call the children is unfortunately driven by our society's rampant genetic essentialism. The majority of Americans still accept the idea that races have a biological essence (and purity).[28] There are no laws that classify individuals by race at either the state or federal level. Race identity is self-reported. Multiracial individuals respond to the genetic essentialism of our society in a variety of ways, which might change over their lifetimes. Some will cling to one ancestry and not the other (often associated with their own appearance). Some will honor both ancestries equally. Joe's experience suggests that the latter course is the more psychologically healthy choice.

WILL MY CHILDREN BE HEALTHY AND GOOD-LOOKING IF I REPRODUCE WITH SOMEONE OUTSIDE MY RACE?

Sorry, but we cannot answer that question. What we can say definitively is that your children's health will not be negatively impacted by the socially defined race of your partner. However, even this statement must be taken with a grain of salt, as we have established that a person's socially defined race might place them in an environment that is not conducive to good health. We explained the impact of the environment on health in chapters 3, 5, and 7. There is no evidence that reproduction with any human from any part of the globe is harmful. Indeed, the general expectation is that reproduction with someone whose ancestry is different from your own should minimize the number of deleterious (bad) genes that are homozygous. In virtually every organism ever studied, an increase in heterozygosity leads to an improvement of the health of the offspring.[29]

Through all the centuries of efforts to document that there might be some biological reason that two individuals of different races should not reproduce, none has ever been found. It is flat-out myth, flat-out wrong to think there is anything biologically problematic. Of course, there is the potential for discrimination and feeling socially ostracized by different communities. This can certainly influence your own and your child's mental health, but it does not have to.

In the nineteenth century, some scientists tried to invent biological reasons for why individuals of different races should not reproduce. They even came up with a popular term for it: *miscegenation,* a term used for the interbreeding of two individuals from different races, which they treated as separate species. The "fertility of the hybrids" was an argument that Charles Darwin used to debunk the idea that human races actually represented separate species.[30] He relied on data that he received from South Carolina slaveholder John Bachman. Bachman's plantation records showed no loss of viability in the children of enslaved African women and white males. We need look no further than Thomas Jefferson, Sally Hemings, and her family to see that reproduction between individuals of different genetic ancestries does not cause biological harm.

Even though polygenism was disproven, the idea that cross-race children would be unhealthy only gained speed. The end of chattel slavery drove this idea. During slavery, the impregnation of Black women by white men was frowned on, but no law prevented it. However, once slavery ended, Southern states rapidly enacted miscegenation statutes, primarily to prevent Black men from impregnating white women. The census categories took care to document the portion of blood by race, mainly to identify those who would be denied their rights under separate and unequal Jim Crow laws. However, it was also thought that Blacks were a reservoir of disease that could be passed into the white community by sexual relations, including the genetic disease sickle cell anemia and the infectious disease syphilis.[31]

In the first half of the twentieth century, scientists rolled up their sleeves to study miscegenation. For example, Charles Davenport, then head of the Cold Spring Harbor lab, studied race crossing and concluded that the result of the race cross was always worse than the parent races. He initially studied immigrant communities in New York (1917) and then headed to Jamaica.[32] He considered race crosses to be disharmonious. Davenport was influential and not alone: he had the weight of the Eugenics Record Office at Cold Spring Harbor behind him.[33]

And this fake science led to laws. In fact, as noted earlier, thirty-three states passed anti-miscegenation laws. An anti-miscegenation amendment to the U.S. Constitution was proposed at various times but never passed into law. The last of these state laws was overthrown by the Supreme Court in 1967 in the case of *Loving v. Virginia*.

The United States was not the only country to legislate against sexual unions between individuals of different races. South Africa did so with national laws in 1949 and 1950. The first eugenics act of the Nazi government was to sterilize the progeny of African soldiers who fathered children with German women after WWI. In 1935, they banned sexual relations between Jews and Germans.[34]

WHY DO SOME PEOPLE SAY THAT INTERRACIAL CHILDREN ARE BETTER LOOKING?

Don't be confused: there has never been a time in America when people weren't having interracial sex and bearing children. However, because of miscegenation laws, it was rare for these couples to marry. Now it is more common. One can legitimately ask whether this trend will have any impact on the future of the race. As we have become increasingly more cosmopolitan, the number of so-called interracial marriages has increased. As noted earlier, President Barack Obama is the result of an interracial marriage. His mother, Ann Dunham (1942–95), who went on to become an anthropologist, was from Wichita, Kansas. His father, Barack Obama, Sr. (1936–82), was from Kenya. His parents met in 1960 while students in a Russian language class at University of Hawaii, Manoa.

The same story, though different, is true for Vice President Kamala Devi Harris. Her mother, Shyamala Gopalan (1938–2009), was a biologist from India. Her father, Donald Harris (1938–), is an economist and emeritus professor at Stanford University. He was an immigrant from Jamaica of African descent. They met at a protest in San Francisco. How crazy are these stories? These days, not very crazy. The rest is history.

These are two public examples of the products of interracial marriage who are now in their fifties. Evidence suggests that what was somewhat unusual in the 1960s has become more common in the 2010s and 2020s. Indeed, when communities and individuals come together, desire happens. However, no one should jump the gun on this trend yet, because it accounts for only about 3 percent of marriages. And, as discussed earlier, who marries whom in interracial marriage dyads is strongly influenced by ongoing patterns of racial inequality.

WHAT DOES "CAUCASIAN" MEAN?

Alan was at home listening to a National Public Radio show during which individuals were being interviewed about their attitudes toward the Black Lives Matter movement. One person identified as Caucasian and another did not but said her husband was Caucasian. Alan's friends often use the term "Caucasian." What do they mean by that?

Doctors and forensic anthropologists use Caucasian. It shows up on police forms and medical intake records. In our research on multiple myeloma, we found that various websites list race as a risk factor. Some specify that being a member of the Black race is such a factor. Blacks have twice the prevalence of multiple myeloma as Caucasians.[35] Caucasians? Who are they?

To the ancient Greeks, "Ethiopian" meant all Africans. However, modern-day Ethiopians are genetically closer to Europeans than to many western African populations.[36] Today, in studies of multiple myeloma, they might be labeled as individuals of African ancestry. Asian was once just Chinese. Negroid (and Negro) and Mongoloid have gone largely out of favor in public discourse, but these nineteenth-century anthropological categories can still be found as search terms in library systems. "Caucasian," by contrast, is still in popular use. Is it not weird how white folks don't even know how to name their own group?

As we noted in earlier chapters, the term "Caucasian" originated in the eighteenth century. German anatomist Johann Blumenbach declared that a skull in his possession from the Caucasus Mountains, an area between the Caspian and Black Seas, was close to the most beautiful that could be found. To Blumenbach, the human skull was created in God's image. Thus, he determined that the particular skull was of the type for individuals like him and in his race. By "type" we mean that it would stand for what God created. All other skulls could then be considered to have changed somehow from the original type. We have previously talked about how eighteenth- and nineteenth-century naturalists generally thought that such change was a degeneration from the ideal type. Blumenbach called the people of this type Caucasians, after the location of the skull. In addition, individuals of other races would need other types, but those could be graded on the degree to which they varied from the skull from the Caucasus, with the further down they were on the typological ladder of life, the greater the variation. It is notable that Georges Cuvier was the only named person in the central plate of Nott and Gliddon's 1857 work, *Indigenous Races of the Earth*.[37]

What constitutes who is Caucasian, or white, has never been determined scientifically. It is the wrong question, because there is nothing there. However, it has been hotly debated for centuries, because the stakes have always been high. Typically, Caucasian included only Europeans, then scientists and others debated the extent of the Caucasian classification. Did it include Middle East- erners? A medical anthropologist colleague of Alan's from Egypt was shocked to find out that when she came to the United States, she became a Caucasian.

Northern Indians were thought by physical anthropologists and linguists to be Caucasians. But that was not so for those who policed the color line. This was well illustrated by the Supreme Court case *United States v. Bha- gat Singh Thind*. In 1923, Thind, an Indian Sikh, sued the United States for citizenship, citing his identity as a "free white person." His evidence for this claim was that anthropological science held that Europeans and Indians shared common descent from Proto-Indo-Europeans. Chief Justice George Sutherland summarized the position of the majority. They held, in effect, that the science of Thind's ancestry did not matter, because he did not meet the common white man's definition of Caucasian.[38]

The Thind case exemplifies the sort of contradiction that stands in the middle of the classification. Blumenbach and later scientists knew that peoples residing in Europe, which had arbitrary boundaries, moved around and intermixed with individuals from Asia and the Middle East. Remember Homer's *Iliad*? Troy was located in modern Turkey. So, would these indi- viduals who mixed with Europeans be Caucasians as well? Ancient DNA (chapter 10) also demonstrates that the modern inhabitants of Europe are descended from farmers who originated in the Middle East.

Blumenbach's fivefold classification of races, which included Caucasian along with Ethiopian, Mongoloid, Malayan, and Native American, had great influence. Many of the descriptions changed. Ethiopian became African and Mongoloid became Asian. But the mythical Caucasian is still kicking around. One of the great myths associated with the term was the idea that all humans are descended from Caucasians. James Cowles Prichard (1786–1848), a pioneer of the typological race concept, believed, as almost everyone did during his time, that humans must have originated in Europe.[39] This commonly agreed-upon notion played a major role in the rejection of Raymond Dart's discovery in 1924 of *Australopithecus africanus* (a close human relative) in Taung, in the northwest province of South Africa.[40] However, as more and more of the oldest fossils of human relatives and anatomically modern humans were discovered in Africa, European and American anthropologists finally had to give up on the myth of Caucasian origins.[41]

It is time to put the term Caucasian to rest. It was never defined, its roots are pure typology, and it invokes a fake science suggesting that the category has some scientific validity. "European" or "European American" is a useful political and social category. One can self-identify as German American or German, Polish American, or Polish, and so on. It is your choice. However, Caucasian serves no purpose but to reify biological race as a typological concept and to obfuscate. Let's make a promise to one another to never again use the term.

ARE "JEWS" A RACE? WHAT ABOUT PEOPLE OF ITALIAN AND IRISH DESCENT?

No. Jews are not a race. Italians and the Irish are not a race. These groups are not races in either the biological or the cultural sense. Italian and Irish are ethnicities in the U.S. context of ethnicities, the lumping together of immigrants from existing countries. Jews are a religious group that also incorporates a couple of ethnicities based on their diasporic locales.

Jews—or, better, Ashkenazi Jews—are a subset of Jews who migrated from Palestine and the Middle East and lived for centuries mostly in isolated villages in central and eastern Europe. Sephardic Jews also left the Middle East and lived in North Africa and the Iberian Peninsula. Most left or converted to Christianity after their expulsion from the peninsula in 1492. A third and smaller branch, the Mizrahi, remained in the Middle East. The diaspora also spun off some smaller and even more isolated branches in modern-day India, Yemen, Ethiopia, and even China. All of these diasporic branches maintained some version of Judaism but came to speak different languages and develop unique customs. They can be considered ethnic groups, in that they share lots of customs and beliefs. The same goes for Irish Catholics and Irish Protestants; they are ethnic groups. Italians are an ethnic group that might be broken down into smaller groups, such as Sicilians and northern Italians.

This view of various Europeans as being ethnic groups is relatively new. Before World War II, it was common to think of all of these groups as races or racial types. Unfortunately, some of this thinking is still promoted in the popular media and among individuals who have not had the opportunity to consider why these groups are not races.

In his influential book, *The Races of Europe* (1899), William Z. Ripley classified Jews, Italians, and the Irish as members of races. The same occurred in 1939 when Harvard physical anthropologist Carleton Coon counted these individuals among his various racial types of Europe. Coon was a walking contradiction in trying to have it both ways: In addition to listing some dozen races of Europe, he kept to the five main quasi-continental races, including the racist notion, outlandish even at the time, that the five races stepped over the threshold to become *Homo sapiens* at different times.[42] Of course, by Coon's theory, Caucasians became human first, with Africans in last place. Following the tradition of Nott, Gliddon, and Ripley, in his books, Coon included lots of side and front view mug shots of individuals to display the different racial types.

Despite the fact that Adolph Hitler knew that race science was bogus, he and Nazi Party officials weaponized race science to support their program of blaming Germany's problems on the Jewish racial element, with the ultimate goal of seizing power in Germany.[43] Jews and Gypsies were studied to determine the faults of these aliens (compared with the mythical Aryans racial types). In a parallel with the ongoing science of Black and brown bodies, Nazi scientists examined bodies and minds of Jews and Gypsies, working hard to pinpoint what was wrong with the Jews and to eradicate the Jewish problem. Jewish blood was a pollutant.

But all of that work is fake science. It is based on ideology and belief. It is wrong in its facts and especially the assumption that Jews, Italians, Irish, and other groups are anything close to closed-breeding populations. Yes, there are some small differences in allele frequencies among these groups (see chapter 2). Yes, some Jews are prone to certain simple Mendelian diseases, such as Tay-Sachs and Gaucher disease. However, through genetic counseling, American Jews have reduced their frequency of Tay-Sachs, to the extent that it now matches other populations of European descent.[44] But the propensity to these diseases constitute small genetic differences in the greater scheme. We have explained in chapters 5 and 9 how human populations (ethnic groups) carry small genetic differences that have little impact on complex traits. So, once again, as Ashley Montagu told the world in 1942, the Jews and European ethnic groups are not races.[45] Some genetic difference? Yes. Race? Absolutely not.

AM I WHITE?

Maybe. The answer depends not just on where you are from and your skin color but also on where you live and especially when you live. The color line between white and various forms of nonwhite has changed over time and from place to place.

First, white is one of those color terms that is used informally to lump together individuals of European ancestry. Of course, nobody is really white, and in fact, some Europeans are closer to beige or light brown. White seems to have become the chosen descriptor of Europeans, as it provided the maximum contrast to Black. The gulf between races might not have seemed so insurmountable if the labels had been beige and brown.

"White," rather than strictly being European, has in some quarters replaced "Caucasian" as a term that has implications for being somewhat genetic and not merely geographic. It is a more social and less anachronistic term than Caucasian for a hard-to-define group of people. Thus, one could be a European—say, an ethnic Pole or an Irish Catholic—in the early twentieth century, but one's whiteness was still questioned. Finally, white has stood for some sort of trans-European white culture.[46] In this regard, white as a designation is fascinating. As most sociologists will tell you, it is really hard to pin down what white culture is, yet it is often fetishized and seen as in need of defending. Sociologists have also begun to examine whiteness as an object of study. This, we feel, is an interesting variation on the study of race and racism, as white is often also unmarked and taken for granted. White individuals often feel that they have no race and, even more often, that they are not constrained or described by their race.

After the horrors of the Holocaust, the distinction among European types began to fade. The United States was to be a melting pot where individuals from different countries and with different religions could go to school and work together, and even love one another. The result was that many European groups became honorary whites, and the category of white expanded beyond Nordic, Anglo-Saxon, and Aryan.[47] However, at the same time, the sharpness of the color line between whites and others became even more significant.

WHAT IS WHITE SUPREMACY?

White supremacy is the belief that white people are superior to those of other races. It claims that because of their intellectual and cultural superiority, whites should be at the top of the social, political, and cultural ladders. White supremacy is the most pernicious myth of the Western world. Behind it is the myth that races are natural and real and that they are ranked hierarchically. Thus, white supremacy is just a new version of eighteenth-century racist thought. It is our time's big lie.

It is hard to quantify how much suffering has been and continues to be caused by adherence to white supremacy. Here, we are not just referring to the more egregious acts, such as the attempted takeover of the U.S. Capital on June 6, 2021, which have resulted from this false belief. The victims of a false ideology are not just persons of color, the intended victims; white supremacy hurts poor and middle-class whites. And frankly, by maintaining a hierarchy supported by a myth rather than merit, it hurts everyone.

White supremacy sits at the core of racism because it reifies that there is such a thing as a white race. Furthermore, it harks back to basic typological thinking in arranging races on a ladder of inferior to superior, with whites at the top as a result of evolution or God's will. We are firm in our conviction that white supremacy is one of the major factors impairing white American Christians from actually following the core tenets of their belief system to "love God and love your neighbor as yourself."[48]

It is also hard to quantify the extent of white supremacist thinking. Only those individuals on the tip of the white supremacist iceberg openly identify as such. However, studies suggest that this is a significant issue in America. Data suggest that virtually every white person who voted for Donald Trump in the 2016 and 2020 elections was, in part, unconsciously motivated by white supremacist thinking. Trump received more than 62 million votes in 2016 and over 74 million votes in 2020, the majority of which were cast by whites. This was indicated by exit polls as well as graphic representations of the percentage of voters by county across the nation.[49] Analysis of why individuals voted for Trump in 2016 at first focused on issues associated with the economy; however, a more detailed study showed that a significant number of Trump voters were primarily motivated by their racial attitudes and support for blocking the immigration of individuals from the Middle East and Central and South America.[50] One of the clearest political links to white supremacy is the belief that America should be a white, Christian country.

Further evidence of the resurgence of white supremacy in America is provided by the analysis of individual and group racial attitudes that are more freely expressed on the internet.[51] In addition, the last decade has witnessed ongoing and intensifying racist incidents on college campuses throughout the country. In November 2019, all fraternity activities at Syracuse University were suspended after video captured fraternity members calling an African American female student by a racial slur.[52] Finally, there is growing evidence that white supremacy has made a resurgence in the military and that little to nothing is being done by commanders to stop it.[53] These examples make clear that white supremacy is more than those committed to white racial hegemony. It is also widespread among those who wish to turn a blind eye to it. And as long as the big lie of white supremacy continues to be oxygenated, we are pessimistic about the future of racial equality and democracy in America.

It is common to think that certain people, such as members of the KKK, are white supremacists. Maybe some older Southerners also harbor white supremacist thoughts. But most whites do not. Then came the riots in Charlottesville in 2017 and the attempted overturn of the 2020 presidential election results and storming of Congress January 6. And the world was exposed to a vast underground of a well-organized white supremacist movement that now also plants its propaganda online.[54]

But those events are just the tip of the iceberg. White supremacy is the air we breathe in the United States. White supremacy is the dominant form of racial thought. It is everywhere. White supremacy is the underlying ideology of racism. We need to understand and face the size and shape of the iceberg of white supremacy.

WHAT IS WHITE NATIONALISM?

White nationalism is a political stance based on white supremacy and strives for either separation of the races or complete control of the government by whites. Despite the cameo appearances of African Americans and

Latinx individuals at the 2020 Republican National Convention, the GOP became the party of white people in the 1960s. Richard Nixon conquered the White House using the same strategy that Donald Trump is attempting to resuscitate for the 2024 election, by playing on white suburban fear of urban racial unrest.

"People are terrified . . . sooner or later the white community is going to retaliate and all the patient work will be undone. And the majority of law-abiding Negroes are going to take the heat."[55] Nixon became the spokesperson for the so-called silent majority, those who wanted a return to normalcy and law and order. However, Nixon was flanked on the right by an even more violent racist, third-party candidate George Wallace, the former anti-integration governor of Alabama.

Prior to Nixon's move to win over Southern whites, that group mostly aligned with the Democratic Party, "Dixiecrats." For example, Franklin Delano Roosevelt struggled to get his New Deal legislation passed with Southern Democrats. They forced him to limit the labor classifications to which the New Deal would apply, specifically removing those categories that were most crucial for keeping African Americans economically subservient and disenfranchised in the South.[56] With Nixon, the positions of the two parties changed over the twentieth century; the Democrats now command more support from nonwhites than do Republicans.

The Southern Poverty Law Center monitors white nationalist groups, or hate groups. They write that white nationalists' core belief is that countries should be organized around individuals identifying as white and that there should be a turn away from multiculturalism.[57] Thus, one objective of the movement is to return to a country that is more completely dominated by whites. Some advocate the removal of nonwhites from America or the creation of white homelands. Of course, the example of Nazi Germany informs us that it is one short step from these objectives to genocide. Nazi ideology tried to eliminate all non-Aryans from Germany. It played on the myth of purity to justify its eugenic sterilization and eventually extermination of non-Aryans.

Short of purging nonwhites, the objective of white nationalism is to restrain the numbers and political power of non-whites. Thus, it is closely aligned with immigration restrictions to maintain a white majority. It also employs a host of other methods to restrict the rights of nonwhites, including purging voter rolls, gerrymandering, and vigilantism. Racial gerrymandering has the effect of diluting the vote of racial minorities. As a result of this practice, the Republican Party now holds some sixteen additional seats in the House of Representatives.

Racial vigilantism has taken a dangerous turn over the last few years. In its most benign form, it consists of white people calling the police to report Black and brown people who are engaged in everyday activities, such as laying their head down in the common room of their college dorm or bird-watching.[58] On the opposite side of privilege, pro-gun whites feel as if it is their obligation to display their AR-15s while shopping at Walmart.[59] In the summer of 2020, a pro-Trump vigilante shot and killed two and severely wounded a third individual in Kenosha, Wisconsin.[60] This occurred following a call for white militants to come to Kenosha to "protect businesses" in danger from anti-police brutality protestors. The protests erupted after an unarmed black man, Jacob Blake, was shot in the back seven times by a police officer. Later that week, a caravan of some six hundred armed pro-Trump demonstrators descended on Portland, Oregon. Their charge was to demonstrate support for the president's "get tough" policy on demonstrators. The demonstrations in Portland were an ongoing response to the murder of George Floyd, reignited by the shooting of Blake. This precipitated an armed clash between the pro-Trump forces and the protesters, leaving one person dead. The dead man, Aaron J. Danielson, belonged to the right-wing group Patriot Prayer.[61] Trump praised the militias for their actions in Portland and defended Kyle Rittenhouse, the young man who allegedly shot and killed demonstrators in Kenosha.[62]

CONCLUSIONS

Naming races carries a lot of weight. Who gets to be counted in the U.S. census and federal programs such as fair housing and employment have consequences for whether and how inequality is tracked. The process of naming is the first step to making something real, to reification. Where society goes from there is up to the individuals in that society. Our past practices are cause for pessimism. Science, as we have seen so often, got involved to try and prove that racial crosses were biologically harmful. These fake studies helped to justify anti-miscegenation laws. In Nazi Germany and South Africa, naming races was the first step to banning interracial sexual relations, and the same was true in the majority of U.S. states.

In the United States today, communities are largely separated by race. Thus, marriages across color lines are still relatively rare, but they are increasing.

And slowly the stigma and stress of being interracial is decreasing. On the other hand, more concerning is the resurgence of white supremacy and white nationalism which threaten to destabilize our democracy. In a worst-case scenario, the United States could return to a de jure white supremacist nation. History is rife with lessons about people who thought the worst-case scenario could never happen to them. We wrote this book because we felt that it was crucial for us to prevent a repetition of these tragedies. What happens from here is up to all of us.

Chapter Eleven

A WORLD WITHOUT RACISM?

In much of the Democratic Party, it's now fashionable to say that America is racist. That is a lie. America is not a racist country.

—NIKKI HALEY, FORMER UN AMBASSADOR, 2020.

What the people want is simple. They want an America as good as its promise.

—BARBARA C. JORDAN, U.S. CONGRESSWOMAN, 1977

Nikki Haley was born Nimrata Nikki Randhawa, the daughter of Sikh parents who immigrated to the United States from Punjabi, India. Like many women, she changed her last name upon marriage. Later in her speech, she described how her parents, and she herself, experienced discrimination as Indian Americans. That seems like an example of systemic racism to us. Democratic vice presidential candidate Kamala Harris's mother was also an Indian American. Aside from their mothers and that they both went into politics, their similarities seem to end there. Harris's mother married a man from Jamaica of African descent. Her parents met at a civil rights rally in Oakland. She strongly identifies with her African heritage, and attended Howard, a historically Black university. Harris, as evidenced by her speaking and writing, is well aware of her personal experiences of racism as well as institutional racism in the United States. She is now the first Afro-Indian and female vice president of the United States.

Whether institutional racism still exists and is a major problem is not something that is decided by anecdotal experiences of famous people. Recently, Mark Robinson, the African American lieutenant governor of North Carolina, assailed the new inclusive K–12 social studies curriculum as "political in nature" and that it unfairly portrayed America as systematically racist.[1] Robinson is known for his support of various conspiracy theories as well as his Islamophobic and homophobic views.

Throughout this book, we have documented the existence of institutional and systemic racism in the United States. It is a topic that appears in virtually every sociology textbook used at the university level. A Google Scholar search on the phrase "institutional racism United States" turns up more than fifty pages of journal articles and books on the topic. Denying the existence of institutional racism in the United States is like denying the existence of gravity or claiming that the Earth is flat. QAnon conspiracy theorists and others who believe that Trump won the 2020 presidential election freely ignore facts and science.

We did not write this book for Nikki Haley, Donald Trump, Mark Robinson, or Stephen Hsu (Asian supremacist, eugenicist, and former vice president of research at Michigan State University[2]). We know that no evidence of any kind exists that can change the views of people who are deeply committed to the notion that America is the land of the "free." In this respect, these individuals resemble special creationists and flat-Earthers. They might well be science deniers and believe what they hear on Alex Jones's InfoWars. Some research suggests that persons displaying extreme racist tendencies have a mental disease.[3]

However, as we have shown, the vast majority of Americans who harbor racial supremacist views are not crazy. Rather, they have absorbed a cultural lie about biological racial differences in athleticism, personality, and intellect. They have bought into the mantra that the position that individuals and groups occupy in society are the result of those people's attributes, rising and falling as if in a color-blind meritocracy, rather than facing up to systems that bias opportunity on the basis of socially defined race.

It is precisely this mantra that needs to be unpacked and relegated to the scrap heap of human history. Socially defined race was invented to support social dominance. It has never served any other purpose. The proof of that claim is found in American history. We began writing this chapter a week after thousands of people gathered at the Lincoln Memorial for the Commitment March of 2020.[4] Despite the fact that the nation was still in the midst of the COVID-19 pandemic, people gathered to address the issue of ongoing police brutality aimed at African Americans. Most of the individuals observed at this gathering were wearing masks, although social distancing was compromised at the large gathering. At that moment, Jacob Blake, paralyzed from the waist down after a Kenosha, Wisconsin, police officer shot him seven times in the back, was shackled to his hospital bed despite not having been charged with a crime.

The following day, then President Trump appeared at a campaign event in New Hampshire. Most of the gathered supporters were not wearing masks and booed when a public address announcement reminded them that wearing a mask was required by Trump's own Coronavirus Task Force and New Hampshire law.[5] None of the unmasked individuals were arrested or shot by police.

An even worse manifestation of Trumpism materialized on January 6, 2021. A mob of thousands, brought to Washington, D.C. by the former president's ongoing denial of the legitimacy of the election of Joe Biden as the new president of the United States, stormed the Capitol building. They had been prompted by Trump's speech that day. During the insurrection, five people died.[6] We might never know the true death toll that resulted from the insurrection, because most of the people in the mob were not wearing masks. Their actions demonstrated that they intended to capture members of Congress and Vice President Mike Pence. We can only guess that if they had captured any of the officials they were looking for, mock trials and public executions would have followed.

These events are powerful evidence that systemic racism and white supremacy continue to be significant problems in America. How do we move forward? Nothing can be achieved without a vision. In this final chapter, we present our vision.

HOW DOES ONE BECOME ANTIRACIST?

We hope that this book has outlined steps toward developing antiracist thinking. This starts by recognizing that one simply needs to apply standard methods of critical thinking to the issue of race and racism. When Joe was dean of general education at North Carolina A&T University, he insisted on implementing a course in critical thinking for every entering student. Alan teaches at Hampshire College and was its dean of natural science and of faculty. Hampshire College focuses on teaching critical thinking and engagement. Critical thinking—being able to gather information and evaluate facts and ideas—should be a standard part of all high school and university curricula. Many programs rely on adding critical thinking competencies as learning objectives across disciplinary coursework. However, we suggest that requiring at least one course in which the focus is on critical thinking (free of specific disciplinary content) is a better way to improve students' skills. A number of excellent texts can be used at the university level for this purpose.[7]

Antiracism is also a moral decision. It starts with a fundamental question regarding whether you believe that we should have a nation that provides everyone with equal opportunity for life, liberty, and the pursuit of happiness. A country is judged by how it treats its poorest and most disadvantaged. This moral sentiment is in our Constitution. We have unfortunately fallen short and seem to be backsliding. It is more common to ignore the poor and foment racism.

Antiracism requires that we all stand together against ongoing racialist and racist practices. America is in the midst of a racial crisis. The police and vigilante killings of Breonna Taylor, George Floyd, and Ahmaud Arbery, as well as the excessive force shooting of Jacob Blake have led to thousands of demonstrations across the country and internationally. Three of these events were caught on video, but they represent only the tip of the iceberg of racial violence in America. They are the canaries in the coal mine of an ongoing and systematized violence that includes mass incarceration, health disparity, hunger, and hopelessness.

With racism and white supremacy operating in so many aspects of American society, one does not have to march in protest demonstrations to make a difference. Indeed, it can be debated that beyond bringing attention to systemic racial violence, protest marches have little impact on making lasting societal change. As well, not everyone has the ability to march and protest. On the other hand, before, during, and after such protests, the most impactful antiracist activities must take place in the communities where we live, work, worship, and play. We discuss what such activity might look like in this chapter.

ISN'T FOCUSING ON RACE MAKING IT EVEN MORE REAL?

We don't believe that we can make race any more real than it is. We could say we wish race was less real, but just wishing it will not make it happen. The simple and more important point is to be clear about what race is and what it is not.

We have shown how socially defined race operates to diminish the life chances of people in virtually every aspect of our society. Social race is real. Racism is real. What an antiracist perspective requires is accurately understanding exactly how racism and white supremacy operate in specific contexts and developing programs of action to redress these injustices.

This is a project that we dare not fail. The United States—and, seemingly, most of the world—is rapidly polarizing around racial justice issues, and this polarization has put democracy at risk. A 2020 survey found that ethnic antagonism strongly erodes Republicans' commitment to democracy. In the survey, 1,151 Republicans and Independents were asked whether they agreed or disagreed with the statement, "The traditional American way of life is disappearing so fast that we may have to use force to save it." The response showed that 50.7 percent strongly agreed or agreed with that statement. An additional 27.7 percent were unsure about whether force would be required.

Another closely related statement was, "A time will come when patriotic Americans will have to take the law into their own hands." Responses showed that 41.3 percent strongly agreed or agreed, and 26.3 percent were unsure.[8] Recently, we have seen examples of right-wing racially inspired vigilante action, as in the case of Kyle Rittenhouse, who killed two protesters and wounded another in Kenosha, Wisconsin (discussed in chapter 10), and of the January 6 mob attacking the Capitol while prominently displaying white supremacist symbols. In this regard, the comment from Tony Caniglia in response to Black Lives Matter protests against his father's restaurant in Omaha, Nebraska, was sobering: "Get rid of the rubber bullets and it's time to go lethal. . . . I promise you that when the first body hits the ground, reality will set in for 95 percent of the rioters and you can use the other 5 percent as target practice."[9] Talking about race does not make it more real, but accurately and effectively talking about race could make it far less dangerous. We can't stop talking about social race until it is insignificant.

WHEN WILL WE KNOW WE HAVE REACHED RACIAL EQUALITY? HOW DO WE GET TO RACIAL EQUALITY? WHAT IS POSTRACIAL?

Answering the "when" question is easier than answering the "how" question. One way we can determine that we have achieved racial equality would be if all the various measures of social achievement—such as the number of billionaires, CEOs of major corporations, those in management positions, university presidents, doctors, lawyers, computer scientists, Senators, and Congresspeople—were represented at the same level in each socially defined race in the general population. Doing so would be good for everyone, because we would draw on the talents of more individuals, and decisions would be made by a more representative and diverse set of these talented folks.

Similarly, we would expect equality in measures of social dysfunction, such as incarceration rates, police killings of unarmed persons, unemployment, exposure to toxic pollutants, domestic violence cases, disability days, and shortened lives. Achieving this equality does not mean increasing any of these measures for whites. Indeed, we would very much like to see a decrease for every group. For example, the white incarceration rate is much too high, and the same is true for white infant mortality and chronic diseases. The wealth and resources of this nation allow us all to have safer, longer, and more productive lives. Racism holds us all back.

We would argue that only the most hardened white supremacists would be against our metric of what racial equality should look like. With this in mind, the real challenge is determining exactly how we can move from our current racially stratified society toward racial justice. Some might think it is counterintuitive that the mistaken idea that we have already achieved a post-racial color-blind society is one of the major impediments to progress. In the opening chapter, we discussed color-blindness, the idea that society should treat everyone the same way. We described how this idea relies on the claim that instead of the ongoing institutional and individual racism that exists in American society, nonracial factors such as market dynamics, naturally occurring phenomena, and the cultural attitudes of minorities themselves are the main causal factors of their social subordination.

One way to gain a sense of the popularity of such ideas is to examine data associated with race relations. For example, figure 11.1 shows the percentage of people who are very satisfied, satisfied, dissatisfied, and very dissatisfied with the position of Blacks and other racial minorities in America between the years 2000–20. The wording of the question is such that it is hard to know what someone might have been thinking when they answered it. One way of thinking might have been motivated by the Obama presidency during the years 2008–16. Yet, during those same years, the killing of unarmed Blacks by police was unabated, mass incarceration continued, and the median wealth of a Black family was only one-tenth that of whites. Thus, the percentage of those who answered "very satisfied" or "satisfied" in this poll (always between 53–45 percent, summed together in fig. 11.1) is disconcerting.

One of the most illuminating studies of current views in the United States concerning race relations addressed the primary reasons that white voters chose Donald Trump over Hillary Clinton in the 2016 presidential election.[10] Since the 1960s, Republicans have generally fared slightly better than Democrats with college-educated whites. The 2016 exit polls showed that Trump had an edge of 4 percentage points in this group over Clinton. However, the gap between the candidates for non-college-educated whites was 40 percent!

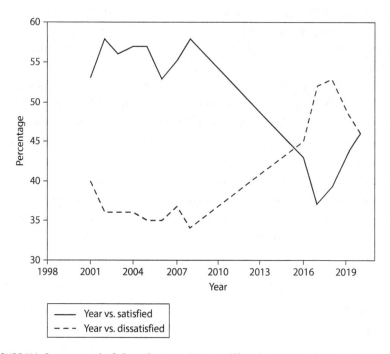

FIGURE 11.1. Responses to the Gallup poll question, "Next, we'd like to know how you feel about the state of the nation in each of the following areas. For each one, please say whether you are—very satisfied, somewhat satisfied, somewhat dissatisfied or very dissatisfied. If you don't have enough information about a particular subject to rate it, just say so. How about—the position of blacks and other racial minorities in the nation?" *Source*: Gallup, "In Depth: Topics A–Z, Race Relations," accessed May 4, 2021, https://news.gallup.com/poll/1687/race-relations.aspx.

These sorts of results make it clear that whether or not America moves in the direction of antiracism is a political issue. Throughout this work, we have demonstrated that biological racism is false. If it is, then white supremacists in America have nothing on which to base their claims. Even if biological races existed within our species, and if some races were smarter than others, there still would be no moral justification for denying all human beings equal treatment under the law. The unanswered question is, how do we move toward a society in which equal treatment under the law is possible?

We argue that answering this question requires dismantling institutions within our society that buttress racial injustice. Many of you who are reading this book might take exception when we begin to outline specific examples of institutions and practices that must be eliminated or significantly altered if we are to ever become a just society. Some of these recommendations exist within the political parties (mainly among Democrats, as the Republicans have so thoroughly hitched their elephant to pull the war chariot of white supremacy). Several go well beyond what most of the Democratic Party would support. Some will begin to make a difference immediately, and others will take longer to have an impact. None of these recommendations is ideal, but achieving them will challenge the way that most people think a society should operate. Indeed, much of what we suggest will have intersectional impacts, impacts on multiple oppressions, such as class, gender, sexual orientation, and racial oppression. In the following sections, we list some actions that we believe will move us in the direction of a less racist and even antiracist society. These recommendations are not all-encompassing or the only ones that we or you could make. Indeed, we encourage you to envision other changes that could help move us toward a more antiracist society.

Voting and Elections

Antiracism starts with fair and equal representation. To that end, we propose the following:

1. All gerrymandering should be immediately struck down. If this simple and obvious action were to be taken, state and federal legislatures would begin to more accurately reflect the racial composition of their constituents.
2. Voter registration should be automatic on every citizen's eighteenth birthday.
3. Everyone should have their voting rights immediately restored on completion of any time served for a crime.
4. All voter suppression tactics (requiring identification, police patrols of voting precincts) should be immediately struck down.
5. Everyone should be able to vote without hardship. Election days should be national holidays. Early voting and absentee voting should be expanded. The number of polling places in low-income neighborhoods should be increased.
6. The Electoral College should be eliminated. It overcounts the votes of states with primarily rural and white majority populations and undercounts those that are more urban with more diverse (Black, brown, yellow) populations.[11] Election of the president should be determined by popular vote. If the Electoral College is retained, "winner takes all" should be eliminated, with electoral votes apportioned based on the percentage of votes each candidate received in each state.
7. Strict limits should be placed on campaign spending. At present, money, more than programs, determines the outcome of elections. Private donations to campaigns

should be eliminated. A fund for local, state, and federal elections should be established, with equal apportionment of funds for all qualified candidates.

8. All campaign ads on television and the internet must be rigorously fact-checked by independent entities before airing. Ads that do not pass fact-checking cannot be aired.

Justice and Policing

The criminal justice system has caused unconscionable harm to poor and especially Black and brown individuals and families. It must be reformed. Specifically, the mission of policing must change from a militarized and often hostile presence in racial minority communities to one that serves and protects all citizens equally. Instead of breaking up peaceful protests or strikes, we propose that the mission of law enforcement be focused— as it always should have been—on protecting all people and all communities from violent crime.

1. Police forces must be diversified to reflect the demography of the communities in which they are assigned to work.

2. Ongoing diversity and equity training must be required within all ranks of the police.[12] It is very important to understand psychological problems and to deal properly with them, and police need more mental health training and more assistance from professionals who are devoted to mental health interventions.

3. Bail reform must be implemented now. The bail system as currently constructed forces hundreds of thousands of poor, predominantly racial minority individuals to sit in jail before they have been convicted of a crime.[13] This has resulted in the creation of a predatory bail bond system that is based on economic means and not on fairness.

4. The death penalty must be eliminated in all states. Racial minorities are differentially impacted by death penalty sentencing. Brian Stevenson and his group have shown that one in nine persons executed was innocent of the crime for which they were sentenced to death.[14] Jennifer Eberhardt and her group have shown that African Americans are more likely to receive the death penalty due to ongoing racist stereotypes concerning their supposed subhuman character.[15]

5. Drugs must be decriminalized and/or legalized. There is no evidence that any socially defined race sells or uses drugs at a substantially different rate.[16] Data from 2015 suggest that about 18 percent of adult whites versus 16 percent of adult

Blacks use drugs. About 2 percent of both groups sell drugs, but Blacks are 6.5 times more likely to be incarcerated at the state level for drug-related offenses.[17] Colorado legalized marijuana in 2012, and by 2020, eleven states had done so. Marijuana is still fully illegal in just eight states. The remaining states have either decriminalized or allowed marijuana use for medical reasons with a prescription. On February 1, 2021, Oregon became the first state to decriminalize small amounts of "hard" drugs.[18] Fully legalizing drugs nationwide would allow taxes to be generated from their sale and distribution. This money can used for funding prevention and addiction treatment programs. Most important, this is money that will not be flowing into organized crime, also reducing the violence associated with the drug trade.

Employment Opportunities: A Massive Job Creation Program

One clear way of improving race relations in this country is to prevent or reduce competition for economic resources. The Reverend Martin Luther King Jr., in his speech in Montgomery, Alabama, pointed out (based on the work of C. Vann Woodward) that segregation gave white supervisors of plantations and factories a backup labor force of poor Blacks. These Black workers could be used to prevent the formation of labor unions or to break up strikes by white workers. This resulted in keeping the wages of Southern whites "almost unbearably low."[19] This has been a consistent theme in the development of capitalism. In the Northern cities, industrialists could always threaten white union workers with employing racial minority strikebreakers.[20]

Competition for jobs has accelerated neo-liberalism. This phase of capitalism has been driven by both the export of labor to foreign countries and the growth of new technology allowing the unparalleled automation of labor. This means that jobs in the manufacturing sector of industrialized nations has been steadily declining. The moniker "rust belt" now describes the cities that were centers of American manufacturing in the mid-twentieth century. The loss of high-paying labor jobs in the American economy is intensifying racial divisions.[21]

Automation via new computer technology has been likened to the steam engine, electricity, and the internal combustion engine. Jobs have disappeared that will never come back in our present economic system. For example, automation has been introduced in American industry to compete with the cost and productivity of labor in underdeveloped nations, such as

Mexico. In 1997, the total manufacturing output per dollar of labor cost for Mexico was 2.7 and only 1.3 for the United States. Via automation, in 2013, that value was 2.9 for Mexico and 2.5 for the United States.[22] So, what happened to displaced American labor? Much of this has moved into the precarious workforce.

Between 1980 and 2004, more than 30 million Americans lost their jobs.[23] Many of these individuals transitioned to long-time unemployed, temporary, and now "gig" economic sectors.[24] People who are employed by companies are protected by minimum wage laws, cannot be dismissed unfairly, can unionize, and receive Social Security and pension benefits. The gig economy does not offer any of that. Even academia has entered the gig world as a result of the unparalleled growth in "adjunct" instructors.[25]

We believe that a false scarcity of jobs severely inhibits real progress toward antiracism in the United States and globally. We are in agreement with the speech by then nominee and now President Joe Biden at the 2020 Democratic National Convention: a commitment must be made to provide meaningful employment for all.

Additionally, insecurity would decrease and development would improve if we could include free early child care and expand Head Start centers, increase the number of public schoolteachers (and their wages), repair infrastructure (such as roads, bridges, and water pipes), and create farms that utilize human labor close to or within inner cities. Indeed, making these changes might have an additional benefit of slowing the production of potentially pandemic-causing viruses.[26] Finally, we support the plan for a "Green New Deal," which promises to create exciting jobs that help to stop global warming and protect our environment. And we need to build a public health system that provides jobs where they are needed and also maximize health, not profits.

Our society needs to envision a new sort of innovation and entrepreneurship. Work need not be full-time. Indeed, most estimates are that efficiencies should lead to the need for less work. But work should be meaningful and well compensated. Instead of driving productivity through decreasing the workforce, innovation should be used to invent new ways of bringing people back into meaningful jobs. The economic advantages of putting people back to work are obvious: more people employed increases the demand for goods and services, which in turn stimulates economic growth. The antiracism benefits are also obvious: reducing economic competition in turn reduces the conflict among socially defined groups.

Education

We wrote this book in hopes of providing an accessible means for individuals to educate themselves and their communities about race and racism. Surveys show that there is still considerable confusion about the meaning of "race" and its influence on society. And indeed, these misunderstandings convinced us of the need for this book. We further propose the following steps to improve education around human variation, race, and racism:

1. K–12 science standards that are associated with human biological variation should be adopted nationwide. Specifically, the standards should address the failure of the biological race concept to describe biological variation within our species. Alan and colleagues Joseph Jones and Maddie Marquez initiated "rethinking race" as a teacher-training program. However, they could not secure funding to continue it. A pilot program headed by Brian Donovan has demonstrated both why this work is so important and how it can be effectively done.[27]

2. K–12 social science and history standards nationwide that accurately address race and racism should be adopted. We must face our history. Specifics include the role of chattel slavery in the formation of the United States, the political fight over slavery as the primary cause for the Civil War, and fuller coverage of reconstruction, the terror of Jim Crow laws, and more recent fights for voting rights and protections. In addition, we advocate for inclusion in K–12 curricula of the racism faced by diverse groups, as illustrated by the Mexican-American War, the wars against American Indian nations, exploitation of Chinese laborers in the building of the Transcontinental Railroad, Japanese internment during World War II, and restrictions on immigration based on faulty racism and eugenics laws of some "European races," such as Jews, Irish, and Poles. Finally, K–12 education should address the expansion of whiteness, further illustrating the role of racism in American history.

3. School funding by real estate values within the district must be eliminated. All public school districts should receive equitable funding. As well, public funding to support charter schools must be eliminated. Charter schools have played a major role in reinstituting school segregation, particularly in the former Confederate states.[28]

4. A kind of "Marshall Plan" for America's public schools is needed. This should include rebuilding and modernizing infrastructure, raising the wages of teachers to competitive levels (particularly in science, technology, engineering, and medicine); expanding the number of teacher aides in the classroom (as part of the jobs

initiatives); expanding the number of counselors and social workers associated with schools; and restoring programs in athletics, arts, and music. This plan will make possible the formation of a labor force for meaningful skilled jobs.

5. Institute truth and reconciliation projects similar to those in many countries, including Germany and South Africa. Susan Neiman's *Learning from the Germans*[29] beautifully illustrates how a nation can face up to its past. These projects should not be seen as scary or anxiety-producing but as opportunities for communities to know and confront their past and move together into a more just future.[30] The pushback by Trump about teaching the history of slavery—specifically the *New York Times* "1619 Project"—shows just how much we need to address, not avoid, our shared history.[31]

Health Care

Access to quality health care ought to be a fundamental human right. This country spends 18–20 percent of its gross domestic product on health care, yet the system is geared to making money rather than eliminating disease and suffering. Moreover, the cost of health care plans is a major contributor to racial health disparity.[32] In addition, in the age of pandemics, the lack of access to health care for the poor threatens everyone, as untreated individuals will act as a reservoir for infectious disease. We propose the following:

1. Adopt a plan of universal health care coverage. The Medicare for All program of Senator Bernie Sanders is a template for the creation of a truly universal health care insurance system by which all Americans will have access to quality and affordable health care.

2. Expand the training and licensing of physicians and physician assistants. There is considerable debate about whether the United States is experiencing a physician shortage.[33] This, of course, is under the current health care model (in which many people do not have access to needed health care, and this shortage differentially impacts racial minorities). It is clear that the number of physicians from racially subordinated groups is inadequate to meet their need.[34] The expanded health care system we envision would require greater numbers of primary care physicians and nurse practitioners. More health care facilities would need to be built in primarily minority and rural communities.

3. Per our discussion in chapter 4, physician training and ongoing professional development programs need curricula that correctly address human biological variation and the causes of disease. Specifically, there must be a focus on

dismantling racial determinist misconceptions concerning disease prevalence[35] and, in its place, an understanding of how racism is a public health problem.

Faith Communities

It is our fervent hope that this book will have some traction within faith communities. The three largest religions in the United States (Christianity, Judaism, and Islam) have tenets that are consistent with antiracism. However, in practice, all have fallen short with regard to their beliefs.

To be clear that what follows is not an attack on Christianity, Joe is a confirmed Episcopalian. He has served on the Racial Justice and Reconciliation Commission of the Episcopal Diocese of North Carolina (RJCR), as well as the Racial Justice Ministry of his own church. That said, white Christians (Evangelicals, Mainline Protestants, and Catholics) have failed the most to face the toxicity of white supremacy in America. Speaking to this claim, Robert Jones commented, "White Christian churches have not just been complacent; they have not only been complicit; rather, as the dominant cultural power in America; they have been responsible for constructing and sustaining a project to protect white supremacy and resist black equality. This project has framed the entire American story."[36]

There is much work to be done in faith communities. Much of this work centers around developing a productive dialog between science and religion. This is not as difficult as it sounds, and tools are available to help facilitate it.[37] Joe serves as the science adviser for both the Chicago and New Brunswick Theological Seminaries (as part of the American Association for the Advancement of Science's Dialogue on Science, Ethics, and Religion). He has discussed science and antiracism before church audiences and for BioLogos.[38]

Following are some things you can do:

1. Start or join an antiracist group in your faith community. There are a number of useful tools. For Christians, Joe recommends Joseph Barndt's *Becoming the Anti-Racist Church*.[39] Although the book is written for Christians, many of its lessons would translate across any faith tradition, or all organizations, for that matter.

2. Engage in real efforts to learn about other faith traditions besides your own. Many faiths are primarily associated with different human populations, such as Islam with Middle Easterners, Europeans, Africans, and East Asians; or Daoism and Buddhism with East Asians. Yet, all traditions include diverse individuals. Learn about the lives of others and their faiths.

INTERSECTIONALITY: HOW DOES ANTIRACISM FIT WITH THE FIGHT AGAINST OTHER FORMS OF OPPRESSION AND SOCIAL INJUSTICE?

We see the fight against racism as hand in hand with the fight against all forms of injustice. The term *intersectionality* is used by scholars to describe the interactions among various forms of social dominance (e.g., classism, racism, sexism, anti-gay bigotry).[40] Women consistently earn less than men, and pay rate also differs by socially defined race. In 2013, female median annual incomes were $46,000 for Asian/Pacific Islanders, $40,000 for whites, $38,000 for people of two or more races, $31,000 for American Indians, and $28,000 for Hispanics.[41]

Homicide rates against transgender individuals also differ by socially defined race.[42] We do not think that we can (or should) make progress against all of these forms of oppression in isolation from one another. For example, one of the drawbacks of the original woman's movement in the United States was its racism. Rampant sexism was one of the most glaring drawbacks of the Civil Rights movement. Finally, we think that it will be difficult or impossible to make lasting change in any form of oppression within the context of capitalism. As discussed earlier, the dynamics of this economic system operate to exacerbate the intersectional antagonisms that exist within our society. Neither do we think that socialism is a guaranteed solution to these social antagonisms. Countries that have made real progress with regard to democratic socialism, such as Norway, have neither large, ethnically diverse populations nor a history of racial slavery. The simple fact is that there is no road map that we can turn to for how to build a more just society. We know the things that we need to dismantle (e.g., classism, racism, sexism, anti-gay bigotry).

Finally, we end with this warning. *If democracy should fall in America, there will be no one to rescue us.* In World War II America, we had the "arsenal of democracy" to provide both materials and people to bring down the fascist and racist Axis forces in Europe and Asia. Joe's dad landed on Utah Beach during the Normandy invasion of 1944. His maternal uncles fought in the Pacific against the empire of Japan. Alan's father was wounded in the war and awarded the Purple Heart. Today, no other nation has the economic means or interest to rescue America from a fascist dictatorship. We are all we have, and therefore the battle against racism is one that we dare not lose. How we do (or do not) engage in this struggle will determine the future of our species and our precious planet.

CONCLUSIONS

Our primary message is that in our species, biological races are a myth. This idea is not supported by modern science. It fails to describe, explain, or tell us how to properly utilize human variation in biomedical research. Yet, race is a social and historical means of classifying and dividing individuals and has many consequences. The belief in biological races has a synergetic relationship to racism. It has provided, and still provides, ideological coverage for racism. Dismantling the myth of race as a biological idea is a first step toward antiracism.

We have tried to clarify what race is and isn't and the connections among race, racism, and human genetic variation. We have endeavored to do so by answering questions that our students, friends, family, and colleagues have asked us, and we have tried to answer them in a straightforward way without skimping on the underlying facts and details. In the following, we present our major takeaways. In this age of elevator pitches and Twitter, we start with some short take-home messages or sound-bites and conclude with some elaborations on the fundamental concepts of race, racism, and human variation.

TEN FACTS

1. Human biological variation is real. It is patterned, important, and a thing of beauty to be celebrated.
2. Race neither describes nor explains human biological variation.

3. Humans do not have biological races.

4. Human biological races are a relatively recent idea that was reified—made real legally and scientifically—to justify racism.

5. Racial classifications developed historically as a politically important means to categorize and divide individuals.

6. Many individuals still believe in the myth of race as being "obviously" biological, in the blood and genes. Race is a powerful illusion. That myth provides ideological justification for systemic racial inequalities.

7. Racism is an ideology that is built on a myth and widely shared, with institutional and structural manifestations.

8. The proof of the impact of racism is found in the data on inequalities in almost all aspects of life, including education, employment, health, and wealth.

9. Race will become less salient when racial ideology is overcome and races reach equality in measures of life such as health and wealth.

10. We cannot have a civil and just society without racial equality. Racial equality will be good for everyone.

HUMAN BIOLOGICAL VARIATION

By this we refer to the pattern of phenotypic and genetic differences and similarities among individuals and groups. Some authors, such as Nicholas Wade,[1] imagine that showing that humans do not have biological races is the same as saying that individual and group genetic variation does not exist or that this variation is always insignificant. Nothing could be further from our position and the facts.

Wade says that the critique of race comes mostly from politically motivated social scientists. He fails to realize that biological anthropologists and evolutionary biologists such as Ashley Montagu, Richard Lewontin, Alan, and Joe have led the critique of biological races. Let's be clear: human variation exists. But race does not describe or explain human variation.

It is true that our overall genetic variation is less than we might have expected. Any individual—of the same or different social race—is surprisingly similar to any other individual on a genetic level. This is mainly because we are a relatively young species and one that has experienced few geographic blocks to the flow of people and their genes.

We are a species with a worldwide distribution. And there is evidence that individuals who occupy different parts of the world differ from one another. These local variations leave some genetic signatures of ancestry, not race.

But these differences are small, found primarily in noncoding genes, and are well explained by geographical distance and specific ancestries. Ancestry is not the same as race.

Biological anthropologists and human population geneticists study the evolutionary forces and mechanisms that produce variations and the various consequences of that variation. As for how variation comes about, most is selectively neutral; thus, groups randomly differ in allele frequencies via genetic drift. The genetic distance between groups is understood by the amount of gene flow between them (isolation by distance). As we showed in chapter 1, the genetic difference between any two groups is highly correlated with the geographic distance between them.

Some variations are adaptive and appear to be the result of natural selection related to environments in different parts of the world. Skin color and sickle cell anemia are two well-studied examples that we previously reviewed. Skin color varies by amount of solar radiation and appears to be a compromise between folate destruction due to too much radiation (most likely to occur in less pigmented skin) and low vitamin D3 (a more likely result for darker skin). Sickle cell trait is a balanced polymorphism. Having one copy of the sickle cell allele is a winning compromise, providing resistance to malaria but not the debilitating consequences of sickle cell disease. It is a genetic compromise.

Other disease risk genes differ in frequency among groups (not races!). However, what is most astonishing is the relatively high amount of variation among individuals within any given group. Because there is so much variation between two individuals in a so-called race, the concept of race ceases to be meaningful. By all objective measures, such as Wright's population subdivision statistic (F_{ST}), humans fail the test of "race-ness."

Finally, individuals vary in simple Mendelian genetic traits such as sickle cell, as well as complex traits such as height, weight, and disease predisposition (determined by many genes). Even the simple traits can be affected by life courses and developmental conditions. The complex traits are just that— complex unfolding of the interactions of multiple genes and environmental conditions over time.

In summary, human biological variation is complex, patterned, important, and incredibly interesting. We are excited to be contributing to the revolution in understanding genomes and genomic differences. Our concern, first, is that we not over-react to genetic information. Indeed, as we explore genomes, what is becoming clearer is just how complex and interactive genes are. And second, we want to make clear (if anything is clear) that human variation is nonracial. *Viva la difference.*

HUMAN BIOLOGICAL VARIATION ≠ RACE

If you've learned one thing from this book, we hope it is this: although human biological and genetic variation is real, humans simply do not divide into races.

The structure of human genetic variation—how it is patterned on a worldwide level—is profoundly nonracial. If humans had races, we would expect clear variations among purported races. But no matter how races are described and delineated, we do not find clear demarcations.

- Human variation is typically continuous from one individual to the next and from one group to the next. For example, skin color around the globe varies slowly from one shade to slightly lighter or darker shades. Allele frequencies also vary slowly from one group to the neighboring groups. Race implies clear breaks or discontinuities, but that is not the truth of human global variation.
- We tend to think of race as identified by phenotypes—outward signs of differences—such as skin color. Biological race implies that skin color is simply the outward manifestation of deeper and more complex traits and abilities. But the truth is that skin color fails to predict almost anything else save the color of one's eyes and hair. This is because traits are largely inherited independently.
- Finally, we now know just how much genetic variation there is among continents, among groups within a continent, and among individuals within any given group. Almost fifty years ago, Richard Lewontin found that only about 6 percent of variation was apportioned to variation among continents and most of the remaining variation was within any local group. Variation among continents and anything we might think of as biological races, Lewontin concluded, are biologically rather meaningless. His results have been corroborated over and over in the last fifty years.
- Variation among groups is mainly a result of geographic distance.
- The average genetic difference between any two individuals has almost nothing to do with race. In fact, two individuals from Africa are on average more different genetically from each other than one of them is from a European or Asian.

RACISM

A main aim of this book is to combat racism as both ideology and institutional practices. Some have argued that racism is a thing of the past and that we are postracial. Others have suggested that racism is a natural part of evolution. Yet others think racism is just a part of the "natural" competition among groups. Some have even argued in the courts and centers of public discourse that favoring individuals of color now overshadows racism against individuals of color. All of these suggestions and ideas are totally unsupported by the facts of inequalities by race in health, wealth, and virtually all important aspects of life.

Ibram X. Kendi makes the important point that racial inequalities in, say, incarceration rates and educational achievements can be explained in only two ways: by genetic differences among races or by unequal historical and continuing social conditions of life. As a naturalist aboard HMS *Beagle*, a young Charles Darwin came to the same conclusion almost two hundred years ago. He wrote, "If the misery of the poor be caused not by the laws of nature, but by our institutions, great is our sin."[2] Indeed, great are our sins.

We have answered questions about why genetic differences among races fail to explain a wide range of inequalities and explain that it is a myth to think that there are genetically based racial differences in attributes such as intelligence, athletic performance, and more. Having shown that genes do not explain racial differences frees us to examine how institutions and systems continue to promote inequalities.

And there are many inequalities. In the United States, there are tremendous differences in wealth and health by class, education, and race. We have tried to show how the race differences in criminal justice and incarceration affect lives and families. Race differences in education, employment, and livable environments perpetuates a permanent caste-like system of inequality. Americans think there is a race difference in wealth but seriously underestimate the size of the gap. There is not a wealth gap; rather, there is wealth hoarding. Everyone suffers without movements toward more economic equality by class and race. Wealthy families become literally fenced and walled off. They live in fear and the nagging reality that their wealth is not sustainable. And, of course, the greater consequence of wealth inequalities is borne by poor families and Black and brown families. Poverty takes away security, leading to stress, and presents a wide range of hindrances to development. Wealth is inherited.

Life expectancy differences between white and Black individuals, now an average of about four years, decreased in the early years of the twenty-first century. That is a great sign, but the difference, on top of class-based inequality, represents an unconscionable loss of opportunity and life. A 2012 paper showed a sixteen-year gap in life expectancy if one combines education and race.[3] And, sadly, now we must add that the racial gap in life expectancy is growing again as a consequence of the grossly unequal impact of COVID-19. A recent estimate is that the reduction in life expectancy at birth in 2020 for whites will be .68 years and 2.10 and 3.05 years for Blacks and Latinos, respectively.[4] Those lost years are of someone's child, spouse, and parent. The loss of opportunity is a drag on society, a drag that everyone pays for.

BEYOND RACIAL THINKING AND RACISM

Racism has its origins in the worldview that races are biologically real and differ in abilities. That view of humankind provided false justifications for enslavement and colonization. It still functions in providing cover for police violence and countless everyday acts that promote the status quo. The most important step to combat racism, therefore, is to expose racial ideology to the light of facts and science.

There are many challenges facing Americans and our globe. We are still in the midst of the COVID-19 pandemic. Climate disruptions are becoming all too evident. A pandemic and climate change are incredibly important global threats. We submit that there is a reason that the United States has failed to come together to fight these threats: racism. Until we change our mindsets to see one another as equals, we are doomed. We will not be able to address existential threats until we address the roots of racism.

Antiracism starts with understanding what race is and isn't.

Antiracism is not just an ethical and scientifically correct position; it is necessary to our survival.

NOTES

PREFACE

1. A. Hassan, "Hate-Crimes Violence Hits 16-Year High, FBI Reports," *New York Times*, November 12, 2019, https://www.nytimes.com/2019/11/12/us/hate-crimes-fbi-report.html.
2. "Antisemitic Incidents Hit All-Time High in 2019," Anti-Defamation League, May 12, 2020, https://www.adl.org/news/press-releases/antisemitic-incidents-hit-all-time-high-in-2019.
3. B. Tatum, *"Why Are All The Black Kids Sitting Together in the Cafeteria?": A Psychologist Explains the Development of Racial Identity* (New York: Basic Books, 1997).

INTRODUCTION

1. Scleroderma Foundation, accessed April 6, 2021, https://www.scleroderma.org/site/DocServer/About_Scleroderma.pdf?docID=660 and Sclero.org, https://sclero.org/scleroderma/causes/genetics/geography.html.
2. "Osteoporosis: Are You at Risk?" WebMD, November 20, 2020, https://www.webmd.com/osteoporosis/guide/osteoporosis-risk-factors.
3. K. Gewertz, "Taxonomist Carl Linnaeus on Show at HMNH," *Harvard Gazette*, November 1, 2007, https://news.harvard.edu/gazette/story/2007/11/taxonomist-carl-linnaeus-on-show-at-hmnh-2/.
4. C. Linnaeus, *Systema Naturae*, 10th ed., 1758.
5. S. Outram, J. L. Graves, J. Powell, et al., "Genes, Race, and Causation: US Public Perspectives About Racial Difference, Race and Social Problems," *Race and Social Problems* 10 (2018): 79–90, https://doi.org/10.1007/s12552-018-9223-7. This study also showed that whites were more likely to believe that race was more biological than social and Blacks were less likely to believe that race was more biological than merely a social classification.

6. M. F. Jacobson, *Whiteness of a Different Color* (Cambridge, MA: Harvard University Press, 1989).

7. We go more deeply into this subject in chapter 3.

8. M. F. Fraga, E. Ballestar, M. F. Paz, et al., "Epigenetic Differences Arise During the Lifetime of Monozygotic Twins," *Proceedings of the National Academy of Sciences of the United States of America* 102, no. 30 (2005): 10604–9, https://doi.org/10.1073/pnas.0500398102.

9. K. J. Wu, "Twins with Covid Help Scientists Untangle the Disease's Genetic Roots," *New York Times*, January 18, 2021, https://www.nytimes.com/2021/01/18/health/covid-twins-genetics.html.

10. C. Darwin, *On the Origin of Species, or the Preservation of Favored Races in the Struggle for Life* (London: John Murray, 1859).

11. N. Jablonski and G. Chapin, "The Evolution of Human Skin Coloration," *Journal of Human Evolution* 39 (2000): 57–106.

12. R. Nielsen, J. M. Akey, M. Jakobsson, J. K. Pritchard, S. Tishkoff, and E. Willerslev, "Tracing the Peopling of the World Through Genomics," *Nature* 541, no. 7637 (2007): 302–10, https://doi.org/10.1038/nature21347.

13. P. M. McDonough, G. J. Duncan, D. Williams, and J. House, "Income Dynamics and Adult Mortality in the United States, 1972–1989," *American Journal of Public Health* 87 (1997): 1476–83.

14. Numerous scholarly studies have demonstrated differential exposure of ethnic minorities to toxic environmental conditions. These studies are reviewed in J. L. Graves, "Looking at the World Through 'Race'-Colored Glasses: The Influence of Ascertainment Bias on Biomedical Research and Practice," in *Mapping "Race": A Critical Reader on Health Disparities Research*, ed. L. E. Gómez and N. López (New Brunswick, NJ: Rutgers University Press, 2003).

15. "On Views of Race and Inequality, Blacks and Whites Are Worlds Apart," Pew Research Center, Social and Demographic Trends, June 27, 2016, https://www.pewsocialtrends.org/2016/06/27/on-views-of-race-and-inequality-blacks-and-whites-are-worlds-apart/.

16. Centers for Disease Control and Prevention, National Center for Health Statistics, "Data Table for Figure 1. Life Expectancy at Birth, by Sex, Race, and Hispanic Origin: United States, 2006–2016," in *Health, United States, 2017*, https://www.cdc.gov/nchs/data/hus/2017/fig01.pdf.

17. This is well explained in J. L. Graves, *The Emperor's New Clothes: Biological Theories of Race at the Millennium* (New Brunswick, NJ: Rutgers University Press, 2005).

18. This perspective is explored in E. Bonilla-Silva, "Rethinking Racism: Toward a Structural Interpretation," *American Sociological Review* 62, no. 3 (1997): 465–80.

19. J. A. Glancy, *Slavery in Early Christianity* (Minneapolis: Fortress, 2006).

20. J. L. Graves, "Genocide (Japanese Occupation)," in *Encyclopedia of Race and Racism*, 2nd ed., ed P. L. Mason (Farmington Hills, MI: Gale-Cengage Learning, 2013).

21. The genocide in Rwanda was a complex and strange application of racialist/racist ideology. For a thorough discussion, see H. H. Hintjens, "Explaining the Genocide in Rwanda," *Journal of Modern African Studies* 37, no. 2 (1991): 241–86.

22. See E. Rueb and D. Taylor, "Tucker Carlson of Fox Falsely Calls White Supremacy a 'Hoax,'" *New York Times*, August 8, 2019, https://www.nytimes.com/2019/08/08/business/media/tucker-carlson-white-supremacy.html.

1. HOW DID RACE BECOME BIOLOGICAL?

1. See the discussion in J. L. Graves, *The Emperor's New Clothes: Biological Theories of Race at the Millennium* (New Brunswick, NJ: Rutgers University Press, 2005), 15–20.

2. J. E. H. Smith, "The Pre-Adamite Controversy and the Problem of Racial Difference in Seventeenth-Century Natural Philosophy," in *Controversies Within the Scientific Revolution*, ed. E. Dascal and N. D. Boantza (Amsterdam: John Benjamins Publishing, 2011).

3. F. Douglass, *Claims of the Negro Ethnologically Considered* (Rochester, NY: Lee, Mann & Co., 1854).

4. Graves, *The Emperor's New Clothes*, chap. 4, "Darwinism Revolutionizes Anthropology," 55–73.

5. A. Smedley, *Race in North America: Origin and Evolution of a Worldview* (Boulder, CO: Westview, 1993).

6. R. McLaughlin, *The Roman Empire and the Silk Routes: The Ancient World Economy and the Empires of Parthia, Central Asia, and Han China* (Barnsley, UK: Pen & Sword History, 2016); and R. C. Redfern, D. R. Gröcke, et al., "Going South of the River: A Multidisciplinary Analysis of Ancestry, Mobility and Diet in a Population from Roman Southwark, London," *Journal of Archaeological Science* 74 (2016): 11–22, http://dx.doi.org/10.1016/j.jas.2016.07.016.

7. Graves, *The Emperor's New Clothes*, 17–20.

8. C. L. Brace, *"Race" Is a Four-Letter Word: The Genesis of the Concept* (New York: Oxford University Press, 2005), 19–20.

9. Brace, *"Race" Is a Four-Letter Word*.

10. I. Hannaford, *Race, The History of an Idea in the West* (Washington, DC: Woodrow Wilson Center Press, 1996).

11. J. Reston, *Dogs of God: Columbus, the Inquisition, and the Defeat of the Moors* (New York: Anchor, 2006).

12. H. Zinn, *A People's History of the United States* (New York: Harper Perennial Modern Classics, 2015). See the chapter titled "Columbus and the Indians."

13. J. Tisby, "What Columbus Really Thought About Native Americans," The Witness, October 8, 2018, https://thewitnessbcc.com/columbus-and-his-fellow-europeans-imported-their-ideas-of-racial-superiority-into-relationships-with-native-americans/.

14. T. Keel, *Divine Variations: How Christian Thought Became Racial Science* (Palo Alto, CA: Stanford University Press, 2019).

15. L. Eiseley, *Darwin's Century: Evolution and the Men Who Discovered It* (New York: Anchor Doubleday, 1958).

16. A. O. Lovejoy, *The Great Chain of Being: A Study of the History of an Idea* (Cambridge, MA: Harvard University Press, 1936).

17. D. B. Domingues da Silva and N. Radburn, eds., *Slave Voyages*, 2008–Present, https://www.slavevoyages.org/.

18. W. M. Wiecek, "The Origins of the Law of Slavery in British North America," *Cardozo Law Review* 17 (1995): 1711.

19. I. F. Gomez, "International Law, Ethno-Cultural Diversity and Indigenous People's Rights," in *Ethno-Cultural Diversity and Human Rights: Challenges and Critiques*, ed. G. Pentassuslia (Leiden: Brill Nijhoff, 2018).

20. D. M. Goldenberg, *The Curse of Ham: Race and Slavery in Early Judaism, Christianity, and Islam* (Princeton, NJ: Princeton University Press, 2003).

21. C. Darwin, *Voyage of the Beagle* (London: Henry Colburn, 1839).

22. E. Mayr, *The Growth of Biological Thought: Diversity, Evolution, and Inheritance* (Cambridge, MA: Belknap Press of Harvard University Press, 1982).

23. S. S. Smith, "An Essay on the Causes of the Variety of Complexion and Figure in the Human Species" (Philadelphia: Printed and sold by Robert Aitken, at Pope's Head, Market Street, MDCCLXXXVII [1787]), http://resource.nlm.nih.gov/2572030R.

24. S. Juengel, "Countenancing History: Mary Wollstonecraft, Samuel Stanhope Smith, and Enlightenment Racial Science," *ELH* 68, no. 4 (2001): 897–927, http://www.jstor.com/stable/30031999.

25. Graves, *The Emperor's New Clothes*, 39, table 3.1.

26. Graves, *The Emperor's New Clothes*, 44, table 3.2.

27. I. Duncan, *Pre-Adamite Man; Or, the Story of Our Old Planet and Its Inhabitants, Told by Scripture and Science* (Edinburgh: W. P. Kennedy, 1860).

28. C. Irmscher, *Louis Agassiz: Creator of American Science* (Boston: Houghton Mifflin Harcourt, 2013); see the chapter entitled "A Pint of Ink."

29. A. Desmond and J. Moore, *Darwin: The Life of a Tormented Evolutionist* (New York: Time Warner, 1991). See also the discussion of Darwin's concern about retaliation from Irmscher, *Louis Agassiz*, 442–43.

30. A. Desmond and J. Moore, *Darwin's Sacred Cause: How a Hatred of Slavery Shaped Darwin's Views on Human Evolution* (Boston: Houghton Mifflin Harcourt, 2009).

31. Mayr, *The Growth of Biological Thought*, 254–63.

32. Graves, *The Emperor's New Clothes*, 44, table 3.1.

33. W. Abraham, "The Life and Times of Anton Wilhelm Amo," *Transactions of the Historical Society of Ghana* 7 (1964): 60–81, http://www.jstor.com/stable/41405765; and B. Brentjes, "Anton Wilhelm Amo, First African Philosopher in European Universities," *Current Anthropology* 16, no. (1975): 443, https://doi.org/10.1086/201577.

34. J. F. Blumenbach, "De Generis Humani Varietate Native Liber" [On the Natural Varieties of Mankind], 1775. In *The Anthropological Treatises of Johann Friedrich Blumenbach and the Inaugural Dissertation of John Hunter*, ed. and trans. T. Bendyshe (London: Longman, Green, Longman, Robers, & Green, 1865), 65–144, http://dx.doi.org/10.5962/bhl.title.50868.

35. Douglass, *Claims of the Negro, Ethnologically Considered.*

36. S. Morton, *Crania Americana, or, A Comparative View of the Skulls of Various Aboriginal Nations of North and South America* (Philadelphia: J. Dobson, 1839), https://collections.nlm.nih.gov/catalog/nlm:nlmuid-60411930R-bk.

37. S. J. Gould, *The Mismeasurement of Man* (New York: Norton, 1980).

38. Irmscher, *Louis Agassiz*, see the chapter titled "A Pint of Ink."

39. R. Bean, "Some Racial Peculiarities of the Negro Brain," *American Journal of Anatomy* 5 (1906): 353–431.

40. Joe summarized the history of racial science in his book *The Emperor's New Clothes*.

41. W. Z. Ripley, *The Races of Europe: A Sociological Study* (New York: D. Appleton, 1899).

42. C. S. Coon, *The Races of Europe* (New York: Macmillan, 1939).

43. R. Bean, "The Nose of the Jew and the Qudratus Mabii Spererioris Muscle," *Anatomical Record* 7 (1913): 47–49.

44. E. O. Manoiloff, "Discernment of Human Races by Blood: Particularly of Russians from Jews," *American Journal of Physical Anthropology* 10 (1927): 11–21.

45. M. F. Jacobson, *Whiteness of a Different Color: European Immigrants and the Alchemy of Race* (Cambridge, MA: Harvard University Press, 1998).

2. EVERYTHING YOU WANTED TO KNOW ABOUT GENETICS AND RACE

1. But even identical twins can vary epigenetically, non-nucleotide changes in DNA, that affect whether DNA is "read" and might eventually code for a protein.
2. Identical twins have identical genomes because the chromosomes they inherited are identical. Identical twins are formed when, after fertilization of an egg by a sperm cell and at the two-cell stage of development, the embryo splits. Each cell now begins development as a separate organism, but both contain an identical set of chromosomes. However, everyone else varies genetically.
3. Ancestry is different from race and should not be confused with the concept of race. We all have ancestries; that is, we come from different places and have parents, grandparents, and ancestors further in our past. Those genetic connections can be traced. This is further discussed in chapter 9.
4. We are also genetically 99 percent the same as chimpanzees, our nearest living nonhuman relatives.
5. R. Redon, S. Ishikawa, K. R. Fitch, L. Feuk, G. H. Perry, T. D. Andrews, H. Fiegler, et al, "Global Variation in Copy Number in the Human Genome," *Nature* 444, no. 7118 (November 23, 2006): 444–54, https://doi.org/10.1038/nature05329.
6. The 1000 Genomes Project Consortium, A. Auton et al., "A Global Reference for Human Genetic Variation," *Nature* 526, no. 7571 (October 1, 2015): 68–74, https://doi.org/10.1038/nature15393.
7. S. Wright, *Evolution and Genetics of Populations: A Treatise in Four Volumes. 4: Variability Within and Among Natural Populations* (Chicago: University of Chicago Press, 1978).
8. This is not an exact science, because populations can be arbitrarily decided on. For example, one could calculate subdivision for a range of ethnic populations within a region of a continent (English, French, Basques), or across regions within a continent (Han Chinese, Japanese, or Mongolians), or among continents (Africa, Asia, Australia, Europe, etc.). Still, F_{ST} does tell us something!
9. S. Garte, "Human Population Genetic Diversity as a Function of SNP Type from Hap-Map Data," *American Journal of Human Biology* 22, no. 3 (2010): 297–300, https://doi.org/10.1002/ajhb.23324.
10. S. Myles, D. Davison, J. Barrett, M. Stoneking, and N. Timpson, "Worldwide Population Differentiation at Disease-associated SNPs," *BMC Medical Genomics* 1, no 22. (2008), https://doi.org/10.1186/1755-8794-1-22.
11. A. R. Templeton, "Genetics and Recent Human Evolution," *Evolution* 61, no. 7 (2007): 1507–19, https://doi.org/10.1111/j.1558-5646.2007.00164.x. However, it is important to avoid making the common error that small amounts of genetic variation within a species necessarily means that the species could not have formed biological races. Unfortunately, this misconception is seen in many popular treatments of biological race. An example is in the description on the race and racial identity section of the National Museum of African American History and Culture's website, "Talking About Race," https://nmaahc.si.edu/learn/talking-about-race/topics/race-and-racial-identity; accessed April 20, 2021.

12. A. Fischer, K. Prüfer, J. M. Good, et al., "Bonobos Fall Within the Genomic Variation of Chimpanzees," *PLoS One* 6, no. 6 (June 29, 2011): e21605, https://doi.org/10.1371/journal.pone.0021605.

13. S. Theodoridis, D. A. Fordham, S. C. Brown, et al., "Evolutionary History and Past Climate Change Shape the Distribution of Genetic Diversity in Terrestrial Mammals," *Nature Communications* 11, no. 2557 (2020), https://doi.org/10.1038/s41467-020-16449-5.

14. D. Thaler, "Why Should Mitochondria Define Species?," *Human Evolution* 33, nos. 1–2 (2018): 1–30, doi:10.14673/HE2018121037.

15. J. Herron and S. Freeman, *Evolutionary Analysis*, 5th ed. (New York: Pearson, 2013).

16. P. Gagneux, M. K. Gonder, T. L. Goldberg, and P. A. Morin, "Gene Flow in Wild Chimpanzee Populations: What Genetic Data Tell Us About Chimpanzee Movement Over Space and Time," *Philosophical Transactions of the Royal Society of London, Series B, Biological Sciences* 356, no. 1410 (2001): 889–97, https://doi.org/10.1098/rstb.2001.0865.

17. R. Kuper and S. Kropelin, "Climate-Controlled Holocene Occupation in the Sahara: Motor of Africa's Evolution," *Science* 313, no. 5788 (2006): 803–807.

18. R. Nielsen, J. M. Akey, M. Jakobsson, et al., "Tracing the Peopling of the World Through Genomics," *Nature* 541, no. 7637 (2017): 302–10, https://www.nature.com/articles/nature21347.

19. Indeed, human activity in reducing the size of other large-bodied mammal populations, their geographical ranges, and their ability to migrate between their locations played a major role in these species' forming biological races.

20. S. Fan, M. E. Hansen, Y. Lo, and S. A. Tishkoff, "Going Global by Adapting Local: A Review of Recent Human Adaptation," *Science* 354, no. 6308 (2016): 54–59, https://doi.org/10.1126/science.aaf5098.

21. N. G. Crawford, D. E. Kelly, M. E. B. Hansen, et al., "Loci Associated with Skin Pigmentation Identified in African Populations," *Science* 358, no. 6365 (2017): eaan8433, https://doi.org/10.1126/science.aan8433. Corrections published December 15, 2017, and January 17, 2020.

22. L. J. Handley, A. Manica, J. Goudet, and F. Balloux, "Going the Distance: Human Population Genetics in a Clinal World," *Trends in Genetics* 23, no. 9 (2007): 432–39, https://doi.org/10.1016/j.tig.2007.07.002.

23. L. Quintana-Murci, O. Semino, H. J. Bandelt, et al., "Genetic Evidence of an Early Exit of *Homo sapiens sapiens* from Africa Through Eastern Africa," *Nature Genetics* 23, no. 4 (1999):437–41.

24. This is what is known as a Neanderthal admixture.

25. T. Günther and M. Jakobsson, "Genes Mirror Migrations and Cultures in Prehistoric Europe: Population Genomic Perspective," *Current Opinions in Genetics & Development* 41 (2016): 115–23, https://doi.org/10.1016/j.gde.2016.09.004.

26. M. Rasmussen, X. Guo, Y. Wang, et al., "An Aboriginal Australian Genome Reveals Separate Human Dispersals Into Asia," *Science* 334, no. 6052 (2011): 94–98, https://doi.org/10.1126/science.1211177.

27. A. S. Malaspinas, M. C. Westaway, C. Muller, et al., "A Genomic History of Aboriginal Australia," *Nature* 538, no. 7624 (2016): 207–14, https://doi.org/10.1038/nature18299.

28. M. Raghavan, M. Steinrücken, K. Harris, et al., "Genomic Evidence for the Pleistocene and Recent Population History of Native Americans," *Science* 349, no. 6250 (2015): aab3884, https://doi.org/10.1126/science.aab3884.

29. A. M. Little, I. Scott, S. Pesoa, et al., "HLA Class I Diversity in Kolla Amerindians," *Human Immunology* 62, no. 2 (2001): 170–79, https://doi.org/10.1016/S0198-8859(00)00248-2.

30. D. Reich, *Who We Are and How We Got Here* (New York: Pantheon, 2018).
31. S. G. Byars and K. Voskarides, "Antagonistic Pleiotropy in Human Disease," *Journal of Molecular Evolution* 88, no. 1 (2020): 12–25, https://doi.org/10.1007/s00239-019-09923-2.
32. M. F. Hammer, A. E. Woerner, F. L. Mendez, et al., "Genetic Evidence for Archaic Admixture in Africa," *Proceedings of the National Academy of Sciences* 108, no. 37 (2011): 15123–28, https://www.pnas.org/content/early/2011/08/29/1109300108.
33. J. K. Pritchard, M. Stephens, and P. Donnelly, "Inference of Population Structure Using Multilocus Genotype Data," *Genetics* 155: 945–59, https://www.genetics.org/content/155/2/945.
34. S. Tishkoff, F. A. Reed, F. R. Friedlaender, et al., "The Genetic Structure and History of Africans and African Americans," *Science* 324: 1035–44, https://doi.org/10.1126/science.1172257.
35. Recent research supports this claim. See T. Günther, C. Valdiosera, H. Malmström, et al., "Ancient Genomes Link Early Farmers from Atapuerca in Spain to Modern-Day Basques," *Proceedings of the National Academy of Sciences* 112, no. 38 (2015): 11917–22, https://doi.org/10.1073/pnas.1509851112; E. R. Jones, G. Gonzalez-Fortes, S. Connell, et al., "Upper Palaeolithic Genomes Reveal Deep Roots of Modern Eurasians," *Nature Communications* 6: 8912, https://doi.org/10.1038/ncomms9912; J. Paganini, L. Abi-Rached, P. Gouret, et al., "HLAIb Worldwide Genetic Diversity: New HLA-H Alleles and Haplotype Structure Description," *Molecular Immunology* 112 (2019): 40–50, https://doi.org/10.1016/j.molimm.2019.04.017; F. A. Reed and S. A. Tishkoff, "African Human Diversity, Origins and Migrations," *Current Opinions in Genetics & Development* 16, no. 6 (2006): 597–605, https://doi.org/10.1016/j.gde.2006.10.008; A. K. Roychoudhury and M. Nei, *Human Polymorphic Genes: World Distribution* (New York: Oxford University Press); O. Thalmann, D. Wegmann, M. Spitzner, et al., "Historical Sampling Reveals Dramatic Demographic Changes in Western Gorilla Populations," *BMC Evolutionary Biology* 11 (2011): 85, https://doi.org/10.1186/1471-2148-11-85; and J. Yang, B. Benyamin, B. P. McEvoy, et al., "Common SNPs Explain a Large Proportion of the Heritability for Human Height," *Nature Genetics* 42, no. 7 (2010): 565–69, https://www.nature.com/articles/ng.608.

3. EVERYTHING YOU WANTED TO KNOW ABOUT RACISM

1. (1963) G. Wallace, "Segregation Now, Segregation Forever," BlackPast, January 22, 2013, https://www.blackpast.org/african-american-history/speeches-african-american-history/1963-george-wallace-segregation-now-segregation-forever/.
2. C. Mills, *The Racial Contract* (Ithaca, NY: Cornell University Press, 1997).
3. I. Wilkerson, *Caste: The Origins of Our Discontents* (New York: Random House, 2020).
4. B. N. Moore and R. Parker, *Critical Thinking*, 9th ed. (Boston: McGraw-Hill, 2009).
5. G. Marcus, *Kluge: The Haphazard Construction of the Human Mind* (Boston: Houghton Mifflin, 2008).
6. W. D. Roth, Ş. Yaylacı, K. Jaffe, and L. Richardson, "Do Genetic Ancestry Tests Increase Racial Essentialism? Findings from a Randomized Controlled Trial," *PLoS One* 15, no. 1 (January 29, 2020): e0227399, https://doi.org/10.1371/journal.pone.0227399.
7. See, for example, I. W. Maina, T. D. Belton, S. Ginzberg, A. Singh, and T. J. Johnson, "A Decade of Studying Implicit Racial/Ethnic Bias in Healthcare Providers Using the Implicit Association Test," *Social Science & Medicine* 199 (2018): 219–29, https://doi.org/10.1016/j.socscimed.2017.05.009.

8. H. D. Fishbein and N. Dess, "An Evolutionary Perspective on Intercultural Conflict: Basic Mechanisms and Implications for Immigration Policy, in *Evolutionary Psychology and Violence: A Primer for Public Policymakers and Public Policy Advocates*, ed. R. W. Bloom and N. Dess (Westport, CT: Praeger, 2003).

9. I. X. Kendi, *How To Be an Antiracist* (New York: One World, 2019).

10. B. Tatum, *Can We Talk About Race?: And Other Conversations in an Era of School Resegregation* (New York: Beacon, 2008).

11. R. DiAngelo, *White Fragility: Why It's So Hard for White People to Talk About Racism* (Boston: Beacon, 2020).

12. F. Hoffman, *Race Traits and Tendencies of the American Negro* (New York: Macmillan, 1896).

13. D. O. Sears, "Symbolic Racism," in *Eliminating Racism: Profiles in Controversy. Perspectives in Social Psychology: A Series of Texts and Monographs*, ed. P. A. Katz and D. A. Taylor (Boston: Springer, 1988).

14. H. Long and A. Van Dam, "The Black-White Economic Divide Is as Wide as It Was in 1968," *Washington Post*, June 4, 2020, https://www.washingtonpost.com/business/2020/06/04/economic-divide-black-households/.

15. D. Muir, "Race: The Mythic Root of Racism," *Sociological Inquiry* 63 (1993): 339–50.

16. K. W. Deutsch, "Anti-Semitic Ideas in the Middle Ages: International Civilizations in Expansion and Conflict," *Journal of the History of Ideas* 6, no. 2 (1945): 239–51.

17. D. Goldenberg, *The Curse of Ham: Race, Slavery, in Early Judaism, Christianity and Islam* (Princeton, NJ: University of Princeton Press, 2003).

18. J. L. Graves, *The Emperor's New Clothes: Biological Theories of Race at the Millennium* (New Brunswick NJ: Rutgers University Press, 2005), 39–44.

19. T. Jefferson, *Notes on the State of Virginia* (Oxford: University of Oxford Text Archive, 1787).

20. C. Darwin, *The Voyage of the Beagle* (New York: Penguin, 1989). First published in 1839.

21. Jefferson's racism is well discussed in Graves, *The Emperor's New Clothes*, 42–43.

22. A. Desmond and J. Moore, *Darwin's Sacred Cause: How a Hatred of Slavery Shaped Darwin's Views on Human Evolution* (Boston: Houghton Mifflin, 2009).

23. C. Darwin, *The Descent of Man and Selection in Relation to Sex* (Princeton, NJ: Princeton University Press, 1981), first published in 1871.

24. R. Moore, "The Dark Side of Creationism," *American Biology Teacher* 66, no. 2 (2004): 85–87.

25. D. R. Sharpe, "Evangelicals, Racism, and the Limits of Social Science Research," *Christian Scholars Review* 33, no. 2 (2004): 237–51.

26. K. Mangan, "A Diversity Director Quit Liberty University. Then He Started 'LUnderground Railroad' for Employees," *Chronicle of Higher Education*, June 9, 2020, https://www.chronicle.com/article/a-diversity-director-quit-liberty-university-then-he-started-lunderground-railroad-for-employees.

27. Project Implicit, Harvard University, accessed April 22, 2021, https://implicit.harvard.edu/implicit/.

28. K. A. Schulman, J. A. Berlin, W. Harless, et al., "The Effect of Race and Sex on Physicians' Recommendations for Cardiac Catheterization," *New England Journal of Medicine* 340, no. 8 (1999): 618–26; slight correction noted in 340, no. 14 (April 8, 1999): 1130.

29. L. A. Penner, J. F. Dovidio, R. Gonzalez, et al., "The Effects of Oncologist Implicit Racial Bias in Racially Discordant Oncology Interactions," *Journal of Clinical Oncology* 34, no. 24 (2016): 2874–80, https://ascopubs.org/doi/abs/10.1200/JCO.2015.66.3658.

30. J. L. Eberhardt, P. G. Davies, V. J. Purdie-Vaughns, and S. L. Johnson, "Looking Death-worthy: Perceived Stereotypicality of Black Defendants Predicts Capital-Sentencing Outcomes," *Psychological Science* 17, no. 5 (May 2006): 383–86, https://doi.org/10.1111/j.1467-9280.2006.01716.x.

31. Donald Trump's relationship with white nationalist Stephen Bannon is described in J. Green, *Devil's Bargain: Steve Bannon, Donald Trump, and the Nationalist Uprising* (New York: Penguin, 2017).

32. For an excellent example of how implicit bias operates in hiring see: D. Pager, *Marked: Race, Crime, and Finding Work in an Era of Mass Incarceration* (Chicago: University of Chicago Press, 2007).

33. Kendi, *How To Be an Antiracist.*

34. Muir, "Race: The Mythic Root of Racism."

35. R. Kennedy, *Nigger: The Strange History of a Troublesome Word* (New York: Vintage, 2003).

36. M. Zimmerman, *Wilhelm Marr: The Patriarch of Anti-Semitism* (New York: Oxford University Press, 1986).

37. The *Protocols of the Elders of Zion* is believed to have been published first in Russia in 1903 and translated into English in 1919. It was instrumental in Nazi anti-Semitic restrictions and eventually the Holocaust.

38. A. Ocalan, *The Sociology of Freedom: Manifesto of the Democratic Civilization* (Chicago: PM Press, 2020).

39. M. F. Jacobson, *Whiteness of a Different Color: European Immigrants and the Alchemy of Race* (Cambridge, MA: Harvard, 1998).

40. E. O. Manoiloff, "Discernment of Human Races by Blood: Particularly of Russians from Jews," *American Journal of Physical Anthropology* 10 (1927): 11–21.

41. R. Bean, "The Nose of the Jew and the Quadratus Labii Superioris Muscle," *Anatomical Record* 7 (1913): 47–49.

42. D. Mogahed and A. Mahmood, *American Muslim Poll: Key Findings 2019*, Institute for Social Policy and Understanding, April 29, 2019, https://www.ispu.org/american-muslim-poll-2019-key-findings/.

43. DiAngelo, *White Fragility.*

44. W. S. Watkins, R. Thara, B. J. Mowry, et al., "Genetic Variation in South Indian Castes: Evidence from Y-Chromosome, Mitochondrial, and Autosomal Polymorphisms," *BMC Genetics* 9 (2008): 86, https://doi.org/10.1186/1471-2156-9-86.

45. G. Myrdal, *An American Dilemma: The Negro Problem and Modern Democracy* (New York: Harper, 1944); M. F. Ashley Montagu, *Man's Most Dangerous Myth: The Fallacy of Race* (New York: Columbia University Press, 1942), xi.

46. E. Bonacich, "Advanced Capitalism and Black/White Race Relations in the United States: A Split Labor Market Interpretation," *American Sociological Review* 41 (1976): 34–51.

47. W. P. O'Hare and B. Curry-White, "Is There a Rural Underclass?," *Population Today* 20 (1992): 6–8.

48. H. Queneau and A. Senc, "On the Structure of US Unemployment Disaggregated by Race, Ethnicity, and Gender," *Economics Letters* 117 (2012): 91–95.

49. B. Hardy and T. Logan, "Race and the Lack of Intergenerational Economic Mobility in the United States," Washington Center for Equitable Growth, February 18, 2020, https://equitablegrowth.org/race-and-the-lack-of-intergenerational-economic-mobility-in-the-united-states/.

50. This is an ongoing discussion in America. See K. Meatto, "Still Separate, Still Unequal: Teaching About School Segregation and Educational Inequality," *New York Times*, May 2, 2019, https://www.nytimes.com/2019/05/02/learning/lesson-plans/stil l-separate-still-unequal-teaching-about-school-segregation-and-educational -inequality.html.

4. WHY DO RACES DIFFER IN DISEASE INCIDENCE?

1. J. Graves, interview by Roland Martin, Unfiltered, March 2020, https://youtu.be /rtRI5hLJ7hc.

2. J. Abbasi, "Taking a Closer Look at COVID-19, Health Inequities, and Racism," *JAMA* 324, no. 5 (2020): 427–29, https://doi.org/10.1001/jama.2020.11672.

3. R. A. Oppel Jr., R. Gebeloff, K. K. Rebecca Lai, W. Wright, and M. Smith, "The Fullest Look Yet at the Racial Inequality of the Corona Virus," *New York Times*, July 5, 2020, https://www.nytimes.com/interactive/2020/07/05/us/coronavirus-latinos-african -americans-cdc-data.html.

4. C. K. Johnson, A. Kastanis, and K. Stafford, "AP Analysis: Racial Disparities Seen in US Vaccination Drive," Associated Press, January 30, 2021, https://apnews.com /article/race-and-ethnicity-health-coronavirus-pandemic-hispanics-d0746b028cf562 31dbcdedaofba24314.

5. CDC Data Tracker, https://covid.cdc.gov/covid-data-tracker/#vaccination-demographics -trends, accessed May 25, 2021.

6. J. L. Graves and M. R. Rose, "Against Racial Medicine," in "Race and Contemporary Medicine: Biological Facts and Fictions," ed. S. Gilman, special issue, *Patterns of Prejudice* 40, nos. 4–5 (2006): 481–93.

7. S. Satel, "I Am a Racially Profiling Doctor," *New York Times*, May 2, 2002, https://www .nytimes.com/2002/05/05/magazine/i-am-a-racially-profiling-doctor.html.

8. K. M. Hoffman, S. Trawalter, J. R. Axt, and M. N. Oliver, "Racial Bias in Pain Assessment and Treatment Recommendations, and False Beliefs About Biological Differences Between Blacks and Whites, *Proceedings of the National Academy of Sciences* 113, no. 16 2016): 4296–301, https://doi.org/10.1073/pnas.1516047113.

9. Hoffman et al., "Racial Bias in Pain Assessment and Treatment Recommendations."

10. For example, see A. H. Goodman, "Why Genes Don't Count (for Racial Differences in Health)," *American Journal of Public Health* 90, no. 11 (2000): 1699–702, https://doi .org/10.2105/AJPH.90.11.1699.

11. J. L. Graves Jr., C. Reiber, A. Thanukos, M. Hurtado, and T. Wolpaw, "Evolutionary Science as a Method to Facilitate Higher Level Thinking and Reasoning in Medical Training," *Evolution, Medicine, & Public Health* 2016, no. 1 (January 2016): 358–68, https://doi.org/10.1093/emph/eow029. This article resulted from a think-tank discussion about medical education that included prominent scientists and physicians.

12. D. R. Snow and K. M. Lanphear, "European Contact and Indian Depopulation in the Northeast: The Timing of the First Epidemics," *Ethnohistory* 35, no. 1 (1988)): 15–33.

13. J. Downs, *Sick from Freedom: African American Illness and Suffering During the Civil War and Reconstruction* (New York: Oxford University Press, 2012).

14. W. E. B. DuBois, *The Philadelphia Negro: A Social Study*, with an introduction by Elijah Anderson (Philadelphia: University of Pennsylvania Press, 1899).

15. A. Nguyen, J. K. David, S. K. Maden, et al., "Human Leukocyte Antigen Susceptibility Map for SARS-CoV-2," *Journal of Virology* 94, no. 13: e00510-20, https://doi.org /10.1128/JVI.00510-20.

16. D. Ellinghaus, F. Degenhardt, L. Bujanda, et al., "Genomewide Association Study of Severe Covid-19 with Respiratory Failure," *New England Journal of Medicine* 383 (2020): 1522–34, https://doi.org/10.1056/NEJMoa2020283.

17. H. Zeberg and S. Pääbo, "The Major Genetic Risk Factor for Severe COVID-19 Is Inherited from Neandertals," *Nature* 587 (2020): 610–12, https://www.nature.com /articles/s41586-020-2818-3.

18. J. G. Read, M. O. Emerson, and A. Tarlov, "Implications of Black Immigrant Health for U.S. Racial Disparities in Health," *Journal of Immigrant Health* 7, no. 3 (2005): 205–12, https://doi.org/10.1007/s10903-005-3677-6; T. G. Hamilton and R. A. Hummer, "Immigration and the Health of U.S. Black Adults: Does Country of Origin Matter?" *Social Science & Medicine* 73, no. 10 (2011): 1551–60, https://doi.org/10.1016/j. socscimed.2011.07.026; and M. J. Brown, S. A. Cohen, and B. Mezuk, "Duration of U.S. Residence and Suicidality Among Racial/Ethnic Minority Immigrants," *Social Psychiatry and Psychiatric Epidemiology* 50, no. 2 (2015): 257–67, https://doi.org/10.1007 /s00127-014-0947-4.

19. R. Nusslock and G. E. Miller, "Early-Life Adversity and Physical and Emotional Health Across the Lifespan: A Neuroimmune Network Hypothesis," *Biological Psychiatry* 80, no. 1 (2016): 23–32, https://doi.org/10.1016/j.biopsych.2015.05.017.

20. L. Dean, *Blood Groups and Red Cell Antigens* (Bethesda, MD: National Center for Biotechnology Information, 2005), https://www.ncbi.nlm.nih.gov/books/NBK2270/.

21. Be the Match, "HLA Basics," https://bethematch.org/transplant-basics/matching-patients -with-donors/how-donors-and-patients-are-matched/hla-basics/.

22. A. R. Sehgal, "The Net Transfer of Transplant Organs Across Race, Sex, Age, and Income," *American Journal of Medicine* 117, no. 9 (2004): 670–75, https://doi.org /10.1016/j.amjmed.2004.05.025.

23. J. Blakely, "Hawaii Attorney General Sues Maker of the Drug Plavix Alleging Unfair and Deceptive Marketing," *Pacific Business News*, March 19, 2014, https://www.bizjournals .com/pacific/news/2014/03/19/hawaii-attorney-general-files-lawsuit-against.html.

24. The most common alleles are *1, *2, *9, *10, and *17. See table 2.6 in S. Stearns and R. Medzhitov, *Evolutionary Medicine* (Sunderland, MA: Sinauer, 2016), 45.

25. J. Kahn, "Race-ing Patents/Patenting Race: An Emerging Political Geography of Intellectual Property in Biotechnology," *Iowa Law Review* 92 (2007): 355–415.

26. K. E. Lohmueller, A. R. Indap, S. Schmidt, et al., "Proportionally More Deleterious Genetic Variation in European than in African Populations," *Nature* 451, no. 7181 (2008): 994–97, https://doi.org/10.1038/nature06611.

27. J. L. Graves, *The Race Myth: Why We Pretend Race Exists in America* (New York: Dutton, 2005).

28. R. Cooper and D. Rotimi, "Hypertension in Blacks," *American Journal of Hypertension* 10, no. (7,pt. 1 (1997)): 804–12, https://doi.org/10.1016/S0895-7061(97)00211-2.

29. See table 2.5 in Stearns and Medzhitov, *Evolutionary Medicine*, 40–41.

30. J. L. Graves, K. L. Hertweck, M. A. Phillips, M. V. Han, L. G. Cabral, T. T. Barber, L. F. Greer, M. F. Burke, L. D. Mueller, and M. R. Rose, "Genomics of Parallel Experimental Evolution in *Drosophila*," *Molecular Biology and Evolution* 34, no. 4 (2017): 831–42, https://doi.org/10.1093/molbev/msw282; and J. L. Graves, A. J. Ewunkem, M. D. Thomas, J. Han, K. L. Rhinehardt, S. Boyd, R. Edmondson, L. Jeffers-Francis, and S.

H. Harrison, "Experimental Evolution of Metal Resistance in Bacteria," in *Evolution in Action: Past, Present, and Future*, ed. Banzhaf et al., 91–106 (Cham, Switzerland: Springer, 2020).

31. J. L. Graves, "Why the Nonexistence of Biological Races Does Not Mean the Nonexistence of Racism," *American Behavioral Scientist* 56, no. 5 (2015): 1474–95, https://doi.org/10.1177/0002764215588810.

32. L. L. Black, R. Johnson, and L. VanHoose, "The Relationship Between Perceived Racism/Discrimination and Health Among Black American Women: A Review of the Literature from 2003 to 2013," *Journal of Racial and Ethnic Health Disparities* 2, no. 1 (2015): 11–20, https://doi.org/10.1007/s40615-014-0043-1.

33. A. Markwick, Z. Ansari, D. Clinch, and J. McNeil, "Perceived Racism May Partially Explain the Gap in Health Between Aboriginal and Non-Aboriginal Victorians: A Cross-Sectional Population Based Study," *SSM—Population Health* 7 (April 19, 2018): 100310, https://doi.org/10.1016/j.ssmph.2018.10.010.

34. N. Snyder-Mackler, J. R. Burger, L. Gaydosh, et al., "Social Determinants of Health and Survival in Humans and Other Animals," *Science* 368, no. 6493 (2020): eaax9553, https://doi.org/10.1126/science.aax9553.

35. J. L. Graves, "Great Is Their Sin: Biological Determinism in the Age of Genomics," *Annals of the American Academy of Political and Social Science* 661, no. 1 (2015): 24–50, https://doi.org/10.1177/0002716215586558.

36. V. Barcelona de Mendoza, Y. Huang, C. A. Crusto, et al., "Perceived Racial Discrimination and DNA Methylation Among African American Women in the InterGEN Study," *Biological Research for Nursing* 20, no. 2 (2018)): 145–52, https://doi.org/10.1177/1099800417748759.

37. S. Oyama and S. F. Terry, "Epigenetics and Racial Health Inequities," *Genetic Testing and Molecular Biomarkers* 20, no. 9 (2016): 483–84, https://doi.org/10.1089/gtmb.2016.29021.sjt.

38. R. Thornton, "Cherokee Population Losses During the Trail of Tears: A New Perspective and a New Estimate," *Ethnohistory* 31, no. 4 (Autumn 1984): 289–300.

39. R. D. Bullard, P. Moha, R. Saha, and B. Wright, *Toxic Wastes and Race at Twenty 1987– 2007: A Report Prepared for the United Church of Christ Justice and Witness Ministries* (Cleveland, OH: United Church of Christ, 2007), https://www.nrdc.org/sites/default/files/toxic-wastes-and-race-at-twenty-1987-2007.pdf.

40. Bullard et al., *Toxic Wastes and Race at Twenty*.

41. Toxic Release Inventory, U.S. Environmental Protection Agency, https://www.epa.gov/toxics-release-inventory-tri-program.

42. J. R. Hipp and C. M. Lakon, "Social Disparities in Health: Disproportionate Toxicity Proximity in Minority Communities Over a Decade," *Health & Place* 16, no. 4 (2010): 674–83, https://doi.org/10.1016/j.healthplace.2010.02.005.

43. M. Trotter, G. E. Broman, and R. R. Peterson, "Densities of Bones of White and Negro Skeletons," *Journal of Bone and Joint Surgery* 42, no. 1 (1960): 50–58.

44. For example, WebMd lists ethnicity as a risk factor for osteoporosis: "Research shows that Caucasian and Asian women are more likely to develop osteoporosis than women of other ethnic backgrounds. Hip fractures are also twice as likely to happen in Caucasian women as in African-American women"; "Osteoporosis: Are You at Risk?," https://www.webmd.com/osteoporosis/guide/osteoporosis-risk-factors. In its entry on risk of osteoporosis, the Mayo Clinic states, "You're at greatest risk of osteoporosis if you're white or of Asian descent"; "Osteoporosis," https://www.mayoclinic.org/diseases-conditions/osteoporosis/symptoms-causes/syc-20351968.

45. In *Savage Africa*, a book published in 1864 in London, W. Winwood Reade, who listed himself as a Fellow of the Geographical and Anthropological Societies of London, wrote, "The skull is extremely thick. If a negro wishes to break a stick, he does not break it across his knee as we do, but across his head. The power of the skull in resisting blows is something marvelous." That same year, Dr. Carl Vogt, professor of natural history at the University of Vienna, wrote that "he [the Negro], like a ram, uses his hard skull in a fight." A year earlier, in the 1863 edition of *Introduction to Anthropology*, Dr. Theodor Waitz wrote, "The skeleton of the Negro is heavier, the bones thicker." Further, he said, "This is especially the case with regard to the skull, which is hard and unusually thick, so that in fighting, Negroes, men and women, butt each other like rams without exhibiting much sensibility."

46. L. Braun, *Breathing Race into the Machine: The Surprising Career of the Spriometer from Plantation to Genetics* (Minneapolis: University of Minnesota Press, 2014).

47. R. T. Jackson et al., "Comparison of Hemoglobin Values in Black and White Male U.S. Military Personnel," *Journal of Nutrition* 113 (1983): 165–71.

48. K. Belson, "Black Former N.F.L. Players Say Racial Bias Skews Concussion Payouts," *New York Times*, August 25, 2020, https://www.nytimes.com/2020/08/25/sports/football/nfl-concussion-racial-bias.html.

49. For more support of these claims, see A. M. Amin, L. Sheau Chin, D. A. Mohamed Noor, et al., "The Effect of CYP2C19 Genetic Polymorphism and Non-Genetic Factors on Clopidogrel Platelets Inhibition in East Asian Coronary Artery Disease Patients," *Thrombosis Research* 158 (2017): 22–24, https://doi.org/10.1016/j.thromres.2017.07.032; G. Hempel, ed., *Handbook of Analytical Separations*, vol. 7: *Methods of Therapeutic Drug Monitoring Including Pharmacogenetics* (Amsterdam: Elsevier Science, 2020), 321–53; J. Kahn, "Race, Pharmacogenomics, and Marketing: Putting BiDil in Context," *American Journal of Bioethics* 6, no. 5 (2006): W1–W5, https://doi.org/10.1080/15265160600755789; C. Roselli, M. D. Chaffin, L. C. Weng, et al., "Multi-Ethnic Genome-Wide Association Study for Atrial Fibrillation," *Nature Genetics* 50, no. 9 (2018): 1225–33, https://doi.org/10.1038/s41588-018-0133-9; B. Séguin, B. Hardy, P. A. Singer, and A. S. Daar, "BiDil: Recontextualizing the Race Debate," *Pharmacogenomics Journal* 8, no. 3 (2008): 169–73, https://doi.org/10.1038/sj.tpj.6500489; D. R. Snow and K. M. Lanphear, "European Contact and Indian Depopulation in the Northeast"; D. Zou and K. L. Goh, "East Asian Perspective on the Interaction Between Proton Pump Inhibitors and Clopidogrel," *Journal of Gastroenterology and Hepatology* 32, no. 6 (2017): 1152–59, https://doi.org/10.1111/jgh.13712; and Z. Shao, L. G. Kyriakopoulou, and S. Ito, "Chapter 14—Pharmacogenomics," in *Handbook of Analytical Separations*, vol. 7: *Methods of Therapeutic Drug Monitoring Including Pharmacogenetics*, ed. G. Hempel (Amsterdam: Elsevier, 2020), https://doi.org/10.1016/B978-0-444-64066-6.00014-9.

5. LIFE HISTORY, AGING, AND MORTALITY

1. S. Stearns and R. Medzhitov, *Evolutionary Medicine* (Sunderland, MA: Sinauer, 2016), 57–64.

2. G. Jasienska, R. G. Bribiescas, A. S. Furberg, S. Helle, and A. Núñez-de la Mora, "Human Reproduction and Health: An Evolutionary Perspective," *Lancet* 390, no. 10093 (2017): 510–20, https://doi.org/10.1016/S0140-6736(17)30573-1.

3. Joe is one of the scientists who helped to establish the scientifically correct modern evolutionary theory of aging. See M. R. Rose, T. Flatt, J. L. Graves, et al., "What is Aging? (opinion)," *Frontiers in Genetics* (July 20, 2012), https://doi.org/10.3389/fgene .2012.00134.

4. M. R. Rose, *Evolutionary Biology of Aging* (New York: Oxford University Press, 1991); J. L. Graves, "General Theories of Aging: Unification and Synthesis," in *Principles of Neural Aging*, ed. S. F. Dani, A. Hori, and G. F. Walter (London: Elsevier, 1997); and M. R. Rose, M. Matos, and H. B. Passananti, *Methuselah Flies: Case History of the Evolution of Aging* (New York: World Scientific, 2004).

5. C. J. Murray, S. C. Kulkarni, C. Michaud, et al., "Eight Americas: Investigating Mortality Disparities Across Races, Counties, and Race-Counties in the United States," *PLoS Medicine* (September 12, 2006): 3:e260, https://doi.org/10.1371/journal.pmed.0030260.

6. S. M. Tajuddin, D. G. Hernandez, B. H. Chen, et al., "Novel Age-Associated DNA Methylation Changes and Epigenetic Age Acceleration in Middle-Aged African Americans and Whites," *Clinical Epigenetics* 11, no. 1 (2019): 119, https://doi.org/10.1186 /s13148-019-0722-1.

7. P. H. Rej, HEAT Steering Committee, C. C. Gravlee, and C. J. Mulligan, "Shortened Telomere Length Is Associated with Unfair Treatment Attributed to Race in African Americans Living in Tallahassee, Florida," *American Journal of Human Biology* 32, no. 3 (2020): e23375, https://doi.org/10.1002/ajhb.23375.

8. R. Coutinho, R. J. David, and J. W. Collins Jr., "Relation of Parental Birth Weights to Infant Birth Weight Among African Americans and Whites in Illinois: A Transgenerational Study," *American Journal of Epidemiology* 146, no. 10 (1997): 804–9, https:// doi.org/10.1093/oxfordjournals.aje.a009197.

9. R. David and J. Collins, "Differing Birthweight Among Infants of U.S. Born Blacks, African-Born Blacks, and U.S. Born Whites," *New England Journal of Medicine* 337, no. 17 (1997): 1209–14. Also see P. D. Gluckman, M. A. Hanson, C. Cooper, and K. L. Thornburg, "Effect of in Utero and Early-Life Conditions on Adult Health and Disease, *New England Journal of Medicine* 359, no. 1 (2008): 61–73, https://doi.org/10.1056 /NEJMra0708473.

10. A. T. Geronimus, "The Weathering Hypothesis and the Health of African American Women and Infants: Evidence and Speculation," *Ethnicity & Disease* 2, no. 3 (1992): 207–21: A. T. Geronimus, "Deep Integration: Letting the Epigenome Out of the Bottle Without Losing Sight of the Structural Origins of Population Health," *American Journal of Public Health* 103, suppl. 1(2013): S56–S63, https://doi.org/10.2105/AJPH.2013.301380.

11. DOHaD was initially known as the "Barker hypothesis" after David Barker. See S. W. Limesand, K. L. Thornburg, and J. E. Harding, "30th Anniversary for the Developmental Origins of Endocrinology," *Journal of Endocrinology* 242, no. 1 (2019): E1–4.

12. Centers for Disease Control and Prevention, National Center for Health Statistics, "Life Expectancy Increases in 2018 as Overdose Deaths Decline Along with Several Leading Causes of Death" (press release), January 30, 2020, https://www.cdc.gov/nchs /pressroom/nchs_press_releases/2020/202001_Mortality.htm.

6. ATHLETICS, BODIES, AND ABILITIES

1. Olympic Diary, "Jamaicans Built to Beat the Rest," Herald.ie, August 19, 2008, https:// www.herald.ie/sport/other-sports/jamaicans-built-to-beat-the-rest-27882487.html.

2. T. Rankinen, M. S. Bray, J. M. Hagberg, et al., "The Human Gene Map for Performance and Health-Related Fitness Phenotypes: The 2005 Update," *Medicine & Science in Sports & Exercise* 38, no. 11 (2006): 1863–88, https://doi.org/10.1249/01.mss .0000233789.01164.4f; and M. S. Bray, J. M. Hagberg, L. Pérusse, et al., "The Human Gene Map for Performance and Health-Related Fitness Phenotypes: The 2006–2007 Update," *Medicine & Science in Sports Exercise* 41, no.1 (2009): 35–73, https://doi .org/10.1249/MSS.0b013e3181844179.

3. J. L. Graves, "Theories of Athletic Performance (General)," in *Encyclopedia of Race and Racism*, 2nd ed., ed. P. L. Mason (Detroit: Macmillan Reference, 2013).

4. The markers were *CREM*, cAMP responsive element modulator, rs1531550 A; *DMD*, encodes the gene dystrophin found in muscle tissue, rs939787 T; *GALNT13*, encodes polypeptide N-acetylgalactosaminyltransferase 13 protein, rs10196189 G; *NFIA-AS1*, encodes the nuclear factor 1 transcription factor, rs1572312 C; *RBFOX1*, RNA binding protein, rs7191721 G; *TSHR*, receptor for thyroid-stimulating hormone, rs7144481 C; I. I. Ahmetov and O. N. Fedotovskaya, "Chapter Six: Current Progress in Sports Genomics," in *Advances in Clinical Chemistry* 70, ed. G. S. Makowski (San Diego: Academic, 2015), 247–314, https://doi.org/10.1016/bs.acc.2015.03.003.

5. The data identified that the polymorphism rs55743914 located in the *PTPRK* gene (which encodes the receptor-type tyrosine protein phosphatase kappa) was the most significant. Seven of the discovered SNPs were also associated with sprint test performance in 126 Polish women, and four were associated with power athlete status in 399 elite Russian athletes. Six SNPs were associated with muscle fiber type in 96 Russian subjects. They also examined genotype distributions and possible associations for sixteen SNPs previously linked with sprint performance. Four of the SNPs (*AGT* rs699, *HSD17B14* rs7247312, *IGF2* rs680, and *IL6* rs1800795) were associated with sprint test performance. In addition, the G alleles of two SNPs in *ADRB2* (rs1042713 and rs1042714) were significantly overrepresented in these players compared with British and European controls; see C. Pickering, B. Suraci, E. A. Semenova, et al., "A Genome-Wide Association Study of Sprint Performance in Elite Youth Football Players," *Journal of Strength and Conditioning Research* 33, no. 9 (2019): 2344–51, https://doi.org/10.1519 /JSC.0000000000003259. Another piece of evidence for the lack of genomic racial differentiation in alleles associated with athletic ability is a GWAS study that found no difference between elite endurance athletes and nonathletes for alleles supposedly related to athletic performance. This study examined 1,520 elite athletes and 2,760 controls from Ethiopia, Kenya, Spain, Poland, Russia, Japan, and Australia. By modern GWAS protocols, this study was underpowered for finding significant genetic differences, because of the small number of people classified as elite athletes. The power to uncover significant differences in the frequency of genetic markers in GWAS is directly influenced by the number of people studied. By modern standards, studies that examine on the order of a million (cases and controls) would be considered very strong, and those examining less than 10,000 (cases and controls) would be considered weak. Alternatively, a GWAS to identify SNPs associated with speed or power in European youth football (soccer) players found twelve SNPs.

6. M. S. Sarzynski, S. Ghosh, and C. Bouchard, "Genomic and Transcriptomic Predictors of Response Levels to Endurance Exercise Training," *Journal of Physiology* 595, no. 9 (2017): 2931–39, https://doi.org/10.1113/JP272559.

7. J. Bale and J. Sung, *Kenyan Running: Movement, Culture, Geography, and Global Change* (London: Frank Cass, 1996).

8. R. Irving and P. Bourne, "The Ecological Model of Sprinting Determinants of Jamaican Athletes," *Austin Emergency Medicine* 2, no.1 (2016): 1009, https://austinpublishing group.com/emergency-medicine/fulltext/aem-v2-id1009.php.

9. All medal counts were gathered from the Wikipedia "All-time Olympic Games medal table," accessed April 29, 2021, https://en.wikipedia.org/wiki/All-time_Olympic_Games _medal_table.

10. H. Kääriäinen, J. Muilu, M. Perola, and K. Kristiansson, "Genetics in an Isolated Population Like Finland: A Different Basis for Genomic Medicine?," *Journal of Community Genetics* 84, no. 4 (2017): 319–26, https://doi.org/10.1007/s12687-017-0318-4.

11. K. Bryc, E. Y. Durand, J. M. Macpherson, D. Reich, and J. L. Mountain, "The Genetic Ancestry of African Americans, Latinos, and European Americans Across the United States," *American Journal of Human Genetics* 96, no. 1 (2015): 37–53, https://doi.org /10.1016/j.ajhg.2014.11.010.

12. J. Benn-Torres, C. Bonilla, C. M. Robbins, et al., "Admixture and Population Stratification in African Caribbean Populations," *Annals of Human Genetics* 72, pt. 1 (2008): 90–98, https://doi.org/10.1111/j.1469-1809.2007.00398.x.

13. J. A. Hawley, M. M. Williams, M. M. Vickovic, and P. J. Handcock, "Muscle Power Predicts Freestyle Swimming Performance," *British Journal of Sports Medicine* 26, no. 3 (1992): 151–55, http://dx.doi.org/10.1136/bjsm.26.3.151; H. C. Emslander, M. Sinaki, J. M. Muhs, et al., "Bone Mass and Muscle Strength in Female College Athletes (Runners and Swimmers)," *Mayo Clinic Proceedings* 73, no. 12 (1998): 1151–60, https://doi.org/10.4065/73.12.1151; and R. Van Schuylenbergh, B. V. Eynde, and P. Hespel, "Prediction of Sprint Triathlon Performance from Laboratory Tests," *European Journal of Applied Physiology* 91, no. 1 (2004): 94–99, https://doi.org/10.1007 /s00421-003-0911-6.

14. M. Lloyd, "Exploring the Racial Disparities in Competitive Swimming," *Swimming World*, February 3, 2016, https://www.swimmingworldmagazine.com/news/exploring-the-racial-disparities-in-competitive-swimming/; S. L. Myers Jr., A. M. Cuesta, and Y. Lai, "Competitive Swimming and Racial Disparities in Drowning," *Review of Black Political Economy* 44, no. 1–2 (2017): 77–97, https://doi.org/10.1007/s12114-017-9248-y.

15. NBC Sports, "US Sets World Record in 4 x 100 Freestyle Relay" (video), July 27, 2019, https://www.youtube.com/watch?v=D7L9T6LuQrA.

16. Joe's history of basketball can be found in J. L. Graves, "Sports (Basketball)" in *Encyclopedia of Race and Racism*, 2nd ed., ed. P. L. Mason (Detroit: Macmillan Reference, 2013).

17. K. Sinsurin, R. Vachalathiti, W. Jalayondeja, and W. Limroongreungrat, "Altered Peak Knee Valgus During Jump-Landing Among Various Directions in Basketball and Volleyball Athletes," *Asian Journal of Sports Medicine* 4, no. 3 (2013): 195–200, https://doi .org/10.5812/asjsm.34258.

18. H. Bryant, *Shut Out: A Story of Race and Baseball in Boston* (New York: Routledge, 2013); and R. K. Barney and D. E. Barney, " 'Get Those Niggers off the Field!': Racial Integration and the Real Curse in the History of the Boston Red Sox," *NINE: A Journal of Baseball History and Culture* 16, no. 1 (2007): 1–9, https://doi.org/10.1353 /nin.2007.0030.

19. G. L. González, "The Stacking of Latinos in Major League Baseball: Does It Matter If a Player Is Drafted?," *Journal of Hispanic Higher Education* 1, no. 4 (2002): 320–28, https://doi.org/10.1177/153819202236976.

20. E. T. Norris, L. Wang, A. B. Conley, et al., "Genetic Ancestry, Admixture and Health Determinants in Latin America," *BMC Genomics* 19, suppl. 8 (2018): 861, https://doi .org/10.1186/s12864-018-5195-7.

21. C. Fortes-Lima, J. Bybjerg-Grauholm, L. C. Marin-Padrón, et al., "Exploring Cuba's Population Structure and Demographic History Using Genome-Wide Data," *Scientific Reports* 8, no. 1 (2018): 11422, https://doi.org/10.1038/s41598-018-29851-3.

22. Wikipedia, "List of World Records in Weightlifting," accessed April 29, 2021, https://en.wikipedia.org/wiki/List_of_world_records_in_Olympic_weightlifting.

23. A. Storey and H. K. Smith, "Unique Aspects of Competitive Weightlifting: Performance, Training and Physiology," *Sports Medicine* 42, no. 9 (2012): 769–90, https://doi.org/10.1007/BF03262294.

24. W. R. Leonard and P. T. Katzmarzyk, "Body Size and Shape: Climatic and Nutritional Influences on Human Body Morphology," in *Human Evolutionary Biology*, ed. M. P. Muehlenbein (Cambridge: Cambridge University Press, 2010).

25. In a recent analysis of literature estimating the heritability of various types of strength (isometric grip, isometric strength, isotonic strength, isokinetic strength, jumping ability, and other measures of power), the heritability (or h^2) for these kinds of strength was determined to be 0.56, 0.49, 0.49, 0.49, 0.55, and 0.51 respectively. All of these estimates of h^2 included their 95 percent confidence intervals. Heritability can vary from 0 to 1.00 and is a measure of how much of a physical trait is determined by genes. This trait is determined by studies of monozygotic twins (those who share all of their genetic information) versus dizygotic twins (who share half of their genetic information), as well as parent–offspring comparisons. The populations examined in this study were primarily European (with some Japanese studies). The h^2 results for these strength traits mean that an individuals' strength is determined as much by their genetic inheritance as the environment in which they live. We also know that due to the relationship between genetic and environmental determinants of physical traits, changing either of these components will result in different estimates of heritability. So, the take-home message is that we cannot easily generalize the results of this study to the rest of the world's populations. See H. Zempo, E. Miyamoto-Mikami, N. Kikuchi, et al., "Heritability Estimates of Muscle Strength-Related Phenotypes: A Systematic Review and Meta-Analysis," *Scandinavian Journal of Medicine and Science in Sports* 17, no. 12 (2017): 1537–46, https://doi.org/10.1111/sms.12804.

26. E. E. Grishina, P. Zmijewski, E. A. Semenova, et al., "Three DNA Polymorphisms Previously Identified as Markers for Handgrip Strength Are Associated with Strength in Weightlifters and Muscle Fiber Hypertrophy," *J Strength Cond Res.* 33, no. 10 (2019): 2602–7, https://doi.org/10.1519/JSC.0000000000003304.

27. M. Collins, K. O'Connell, and M. Posthumus, "Genetics of Musculoskeletal Exercise-Related Phenotypes," in *Genetics and Sports*, Medicine and Sports Science 61, 2nd ed., ed. M. Posthumus and M. Collins (Basel: Karger, 2016); also see M. J. Alter, *Science of Flexibility*, 3rd ed. (Champaign, IL: Human Kinetics, 2004).

28. D. A. Nelson, G. Jacobsen, D. A. Barondess, and A. M. Parfitt, "Ethnic Differences in Regional Bone Density, Hip Axis Length, and Lifestyle Variables Among Healthy Black and White Men," *Journal of Bone and Mineral Research* 10, no. 5 (1995): 782–87, https://doi.org/10.1002/jbmr.5650100515.

29. C. Mukwasi, L. Stranix Chibanda, J. Banhwa, and J. A. Shepherd, "US White and Black Women Do Not Represent the Bone Mineral Density of Sub-Saharan Black Women," *Journal of Clinical Densitometry* 18, no. 4 (2015): 525–32, https://www.sciencedirect.com/science/article/abs/pii/S1094695015001213?via%3Dihub.

30. N. Lovšin, J. Zupan, and J. Marc, "Genetic Effects on Bone Health," *Current Opinion in Clinical Nutrition and Metabolic Care* 21, no. 4 (2018): 233–39, https://doi.org/10.1097/MCO.0000000000000482.

31. L. Michou, "Epigenetics of Bone Diseases," *Joint Bone Spine* 85, no. 6 (2018): 701–7, https://doi.org/10.1016/j.jbspin.2017.12.003.

32. Supporting references for the chapter include T. Rankinen, N. Fuku, B. Wolfarth, et al., "No Evidence of a Common DNA Variant Profile Specific to World Class Endurance Athletes," *PLoS One* 11, no. 1 (January 29, 2016): e0147330, https://doi.org/10.1371/journal.pone.0147330; and A. Vander, J. Sherman, and D. Luciano, *Human Physiology: The Mechanisms of Body Function*, 8th ed. (Boston: McGraw Hill, 2001).

7. INTELLIGENCE, BRAINS, AND BEHAVIORS

1. R. Sternberg, "Intelligence: State of the Art," *Dialogues Clinical Neuroscience* 14, no. 1 (2012): 19–27.

2. H. H. Goddard, "Mental Tests and the Immigrant," *Journal of Delinquency* 5, no. 2 (1917).

3. See Lothrop Stoddard report: "Europe as an Emigrant Exporting Continent and the United States as an Emigrant Receiving Nation," Hearings, Committee on Immigration and Naturalization, March 8, 1924.

4. A. J. Onwuegbuzie and C. E. Daley, "Racial Difference in IQ Revisited: A Synthesis of Nearly a Century of Research," *Journal Black Psychology* 27, no. 2 (2001): 209–20.

5. R. Herrnstein and C. R. Murray, *The Bell Curve: Intelligence and Class Structure in American Life* (New York: Free Press, 1994).

6. M. Hout, "Test Scores, Education, and Poverty," in *Race and Intelligence: Separating Science from Myth*, ed. J. Fish (Mahwah, NJ: Lawrence Erlbaum, 2002).

7. See table 7.2 in R. Lewis, *Human Genetics: Concepts and Applications*, 5th ed. (Boston: McGraw-Hill, 2003).

8. C. Murray, *Human Diversity: The Biology of Gender, Race, and Class* (New York: Twelve, 2020).

9. R. Plomin and S. von Stumm, "The New Genetics of Intelligence," *Nature Reviews Genetics* 19, no. 3 (2018): 148–59, https://doi.org/10.1038/nrg.2017.104.

10. G. Davies, A. Tenesa, A. Payton, et al., "Genome-Wide Association Studies Establish That Human Intelligence Is Highly Heritable and Polygenic," *Molecular Psychiatry* 16, no. 10 (2011): 996–1005, https://doi.org/10.1038/mp.2011.85.

11. S. L. Spain, I. Pedroso, N. Kadeva, et al., "A Genome-Wide Analysis of Putative Functional and Exonic Variation Associated with Extremely High Intelligence," *Molecular Psychiatry* 21, no. 8 (2016): 1145–51, https://doi.org/10.1038/mp.2015.108.

12. A. Okbay, J. P. Beauchamp, M. A. Fontana, et al., "Genome-Wide Association Study Identifies 74 Loci Associated with Educational Attainment," *Nature* 553, no. 7694 (2016): 539–42, https://doi.org/10.1038/nature17671; and A. Kong, M. L. Frigge, G. Thorleifsson, et al., "Selection Against Variants in the Genome Associated with Educational Attainment," *Proceedings of the National Academy of Sciences* 114, no. 5 (2017): E727-32, https://doi.org/10.1073/pnas.1612113114.

13. J. J. Lee, R. Wedow, A. Oakbay, et al., "Gene Discovery and Polygenic Prediction from a Genome-Wide Association Study of Educational Attainment in 1.1 Million Individuals," *Nature Genetics* 50, no. 8 (2018): 1112–21, https://doi.org/10.1038/s41588-018-0147-3.

14. H. Zhang, H. Zhou, T. Lencz, et al., "Genome-Wide Association Study of Cognitive Flexibility Assessed by the Wisconsin Card Sorting Test," *American Journal of Medical Genetics, Part B: Neuropsychiatriatic Genetics* 117, no. 5 (2018): 511–19, https://doi

.org/10.1002/ajmg.b.32642. This study found two significant SNPs associated with cognitive flexibility in African Americans (rs7165213 and rs35633795). The SNPs were located downstream of a noncoding gene, *LOC101927286*.

15. D. Reich, *Who We Are and How We Got Here: Ancient DNA and the New Science of the Human Past* (New York: Doubleday, 2018), 255:

> The indefensibility of the orthodoxy is obvious at almost every turn. In 2016, I attended a lecture on race and genetics by the biologist Joseph L. Graves Jr. at the Peabody Museum of Archaeology and Ethnography at Harvard. At one point, Graves compared the approximately five mutations known to have large effects on skin pigmentation and that are obviously different in frequency across populations to the more than ten thousand genes known to be active in human brains. He argued that in contrast to pigmentation genes, the patterns at genes active in the brain would surely average out over so many locations, with some mutations nudging cognitive and behavioral traits in one direction and some pushing in the other direction. But that argument doesn't work.

16. J. J. McKee, F. E. Poirer, and S. McGraw, *Understanding Human Eolution*, 5th ed. (Upper Saddle River, NJ: Prentice/Pearson, 2005).

17. R. Boyd and J. B. Silk, *How Humans Evolved*, 3rd ed. (New York: Norton, 2003).

18. J. L. Graves, "What a Tangled Web He Weaves: Race, Reproductive Strategies, and Rushton's Life History Theory," *Anthropological Theory* 2, no. 2 (2002): 131–54; and J. L. Graves, "In Defense of the 'Orthodoxy': Who We Really Are and Why Some Folks Have a Problem With It," in *Critical Race Theory in the Academy*, ed. V. L. Farmer and E. S. W. Farmer (Charlotte, NC: Information Age, 2020).

19. G. Zhang, L. J. Muglia, R. Chakraborty, J. M. Akey, and S. M. Williams, "Signatures of Natural Selection on Genetic Variants Affecting Complex Human Traits," *Applied & Translational Genomics* 2 (2013): 78–94.

20. P. C. Sabeti, S. F. Schaffner, B. Fry, et al., "Positive Natural Selection in the Human Lineage," *Science* 312, no. 5780 (2006): 1614–20.

21. See J. L. Graves Jr., K. L. Hertweck, M. A. Phillips, et al., "Genomics of Parallel Experimental Evolution in *Drosophila*," *Molecular Biology and Evolution* 34, no. 4 (2017): 831–42, https://academic.oup.com/mbe/article/34/4/831/2897202; J. K. Pritchard, J. K. Pickrell, and G. Coop, "The Genetics of Human Adaptation: Hard Sweeps, Soft Sweeps, and Polygenic Adaptation," *Current Biology* 20, no. 4 (2010): R208—15, https://doi.org/10.1016/j.cub.2009.11.055; and Zhang et al., "Signatures of Natural Selection on Genetic Variants."

22. Zhang et al., "Signatures of Natural Selection on Genetic Variants."

23. This is well summarized in J. L. Graves, *The Emperor's New Clothes: Biological Theories of Race at the Millennium* (New Brunswick, NJ: Rutgers University Press, 2005); and J. L. Graves, *The Race Myth: Why We Pretend Race Exists in America* (New York: Dutton, 2005).

24. The basic principles of equalizing environment for the estimation of genetic contributions to complex traits is explained in the classic work, by D. S. Falconer and T. MacKay, *Introduction to Quantitative Genetics*, 4th ed. (Essex, UK: Longman, 1996).

25. The genes associated with these traits are *AVPR1* (receptor for arginine vasopressin), *SLC6A4* (membrane protein that transports the neurotransmitter serotonin), *COMT* (encodes catechol-O-methyltransferase), *DRD2* (encodes the D2 subtype of

the dopamine receptor), and *TPH1* (encodes the protein that is involved in step 1 of the sub-pathway that synthesizes serotonin from the amino acid L-tryptophan); see B. Gingras, H. Honing, I. Peretz, L. J. Trainor, and S. E. Fisher, "Defining the Biological Bases of Individual Differences in Musicality," *Philosophical Transactions of the Royal Society B Biological Sciences* 370, no. 1664 (2015): 20140092, https://royalsocietypublishing .org/doi/10.1098/rstb.2014.009.

26. The chromosome positions were 4p14-13, 4p12-q12, 4q22, 8q13-21, 18q12-21, 16q21-22, and 22q11. Specific genes were *GPR98* (encodes a G protein coupled superfamily), *USH2A* (encodes usherin, a protein that is an important component of basement membranes that support cells and separate tissues; both genes are involved in auditory reception), *GRIN2B* (encodes the protein Glu2B, which works in nerve cells, *IL1A* (interleukin 1 alpha releases by macrophages and stimulates production of cells from the thymus gland), *IL1B* (a potent inflammatory cytokine), and *RAPGEF5* (guanine nucleotide exchange factor for cognition, memory). See X. Liu, C. Kanduri, J. Oikkonen, et al., "Detecting Signatures of Positive Selection Associated with Musical Aptitude in the Human Genome," *Scientific Reports* 6 (2016): 21198, https://www .nature.com/articles/srep21198.

27. J. Oikkonen, Y. Huang, P. Onkamo, et al., "A Genome-Wide Linkage and Association Study of Musical Aptitude Identifies Loci Containing Genes Related to Inner Ear Development and Neurocognitive Functions," *Molecular Psychiatry* 20, no. 2 (2015): 275–82, https://www.nature.com/articles/mp20148.

28. Oikkonen et al., "A Genome-Wide Linkage and Association Study of Musical Aptitude."

29. Available through the National Center Biological Information (NCBI), https://www .ncbi.nlm.nih.gov/snp/?cmd=search.

30. S. Sanchez-Roige, J. C. Gray, J. MacKillop, C. H. Chen, and A. A. Palmer, "The Genetics of Human Personality," *Genes, Brain and Behavior* 17, no. 3 (2018): e12439, https:// onlinelibrary.wiley.com/doi/full/10.1111/gbb.12439; and H. J. Foldes, E. E. Dueher, and D. S. Ones, "Group Differences in Personality: Meta-Analyses Comparing Five US Racial Groups," *Personnel Psychology* 61 (2008): 579–616, https://onlinelibrary.wiley .com/doi/abs/10.1111/j.1744-6570.2008.00123.x.

31. Foldes et al., "Group Differences in Personality."

32. J. Sidanius and F. Pratto, *Social Dominance: An Intergroup Theory of Social Heirarchy and Oppression* (Cambridge, UK: Cambridge University Press, 1999).

33. F. Pratto, J. L. Liu, S. Levin, et al., "Social Dominance Orientation and the Legitimatization of Inequality Across Cultures," *Journal of Cross-Cultural Psychology* 31, no. 3 (2000): 369–409, https://journals.sagepub.com/doi/abs/10.1177/0022022100031003005.

34. B. H. Kim, H. N. Kim, S. J. Roh, et al., "GWA Meta-Analysis of Personality in Korean Cohorts," *Journal of Human Genetics* 60, no. 8 (2015): 455–60, https://www.nature.com /articles/jhg201552.

35. D. G. Amen, *The End of Mental Illness: How Neuroscience Is Transforming Psychiatry and Helping to Prevent or Reverse Mood and Anxiety Disorders, ADHD, Addictions, PTSD, Psychosis, Personality Disorders, and More* (Carol Stream, IL: Tyndale Momentum, 2020).

36. R. Nesse, *Good Reasons for Bad Feelings* (New York: Dutton, 2019).

37. T. A. LaVeist, *Minority Populations and Health: An Introduction to Health Disparities in the United States* (San Francisco: Jossey-Bass, 2005).

38. S. J. Bartels and P. DiMilia, "Why Serious Mental Illness Should Be Designated a Health Disparity and the Paradox of Ethnicity," *Lancet Psychiatry* 4, no. 5 (2017):

351–52, https://www.thelancet.com/journals/lanpsy/article/PIIS2215-0366(17)30111-6 /fulltext.

39. D. M. Barnes and L. M. Bates, "Do Racial Patterns in Psychological Distress Shed Light on the Black-White Depression Paradox? A Systematic Review," *Social Psychiatry and Psychiatric Epidemiology* 52, no. 8 (2017): 913–28, https://link.springer.com /article/10.1007/s00127-017-1394-9.

40. Barnes and Bates, "Do Racial Patterns in Psychological Distress Shed Light on the Black-White Depression Paradox?"

41. S. Lipsky, M. A. Kernic, Q. Qiu, and D. S. Hasin, "Traumatic Events Associated with Posttraumatic Stress Disorder: The Role of Race/Ethnicity and Depression," *Violence Against Women* 22, no. 9 (2016): 1055–74, https://journals.sagepub.com/doi/10.1177 /1077801215617553.

42. Lipsky et al., "Traumatic Events Associated with Posttraumatic Stress Disorder."

43. M. Bresnahan, M. D. Begg, A. Brown, et al., "Race and Risk of Schizophrenia in a US Birth Cohort: Another Example of Health Disparity?," *International Journal of Epidemiology* 36, no. 4 (2007): 751–58, https://academic.oup.com/ije/article/36/4/751/665657.

44. K. Plowden, "Schizophrenia Spectrum Disorder Disparity Among African-Americans," *Journal of the National Black Nurses Association* 5*30, no. 1 (2019): 14–20.

45. E. F. Torrey and R. H. Yolken, "Schizophrenia and Infections: The Eyes Have It," *Schizophrenia Bulletin* 43, no. 2 (2017): 247–52, https://doi.org/10.1093/schbul/sbw113; and J. Janoutová, P. Janácková, O. Serý, et al., "Epidemiology and Risk Factors of Schizophrenia," *Neuroendocrinology Letters* 37, no.1 (2016): 1–8.

46. B. Crespi, P. Stead, and M. Elliot, "Evolution in Health and Medicine Sackler Colloquium: Comparative Genomics of Autism and Schizophrenia," *Proceedings of the National Academy of Science* 107, suppl. 1 (2010): 1736–41, https://www.pnas.org/content /early/2010/01/14/0906080106.

47. S. G. Byars, S. C. Stearns, and J. J. Boomsma, "Opposite Risk Patterns for Autism and Schizophrenia Are Associated with Normal Variation in Birth Size: Phenotypic Support for Hypothesized Diametric Gene-Dosage Effects," *Proceedings of the Royal Society B: Biological Sciences* 281, no. 1794 (2014): 20140604, https://royalsocietypublishing .org/doi/10.1098/rspb.2014.0604.

48. L. K. Pries, S. Gülöksüz, and G. Kenis, "DNA Methylation in Schizophrenia," in Delgado-Morales R. (eds) *Neuroepigenomics in Aging and Disease*, Advances in Experimental Medicine and Biology 978, ed. R. Delgado-Morales (Cham, Switzerland: Springer), https://doi.org/10.1007/978-3-319-53889-1_12.

49. M. W. Tremblay and Y. H. Jiang, "DNA Methylation and Susceptibility to Autism Spectrum Disorder," *Annual Review of Medicine* 70 (2019): 151–66, https://www.annual reviews.org/doi/10.1146/annurev-med-120417-091431.

50. M. Lam, C.-Y. Chen, Z. Li, et al., "Comparative Genetic Architectures of Schizophrenia in East Asian and European Populations," *Nature Genetics* 51, no. 12 (2019): 1670–78, https://www.nature.com/articles/s41588-019-0512-x.

51. T. R. de Candia, S. H. Lee, J. Yang, et al., "Additive Genetic Variation in Schizophrenia Risk Is Shared by Populations of African and European Descent," *American Journal of Human Genetics* 93, no. 3 (2013): 463–70, https://www.sciencedirect.com/science /article/pii/S000292971300325X.

52. B. H. Hidaka, "Depression as a Disease of Modernity: Explanations for Increasing Prevalence," *Journal of Affective Disorders* 140, no. 3 (2012): 205–14, https://doi .org/10.1016/j.jad.2011.12.036.

53. C. Boscher, L. Arnold, A. Lange, and B. Szagun, "Die Last der Ungerechtigkeit. Eine Längsschnittanalyse auf Basis des SOEPs zum Einfluss subjektiv wahrgenommener Einkommensgerechtigkeit auf das Risiko einer stressassoziierten Erkrankung [The load of injustice: A longitudinal analysis of the impact of subjectively perceived income injustice on the risk of stress-associated diseases based on the German Socio-Economic Panel Study]," *Gesundheitswesen* 80, S 02 (2018): S71-S79, http://doi:10.1055/s-0043-107876.

54. J. P. Rushton, *Race, Evolution, and Behavior: A Life History Perspective* (New Brunswick, NJ: Transaction, 1994).

55. A. Chase, *The Legacy of Malthus: The Social Costs of the New Scientific Racism* (Champagne-Urbana, IL: University of Illinois Press, 1980).

56. L. Delevingne, "The Decade's 10 Biggest Financial Crimes," Business Insider, December 21, 2009, https://www.businessinsider.com/the-decades-10-biggest-financial-crimes-2009-12#7-scott-rothstein-4.

57. The home page for *Criminal Justice and Behavior* is found at https://journals.sagepub.com/home/cjb.

58. A. M. Gard, H. L. Dotterer, and L. W. Hyde, "Genetic Influences on Antisocial Behavior: Recent Advances and Future Directions," *Current Opinion in Psychology* 27 (2019): 46–55, https://www.sciencedirect.com/science/article/pii/S2352250X18300952?via%3Dihub. Two of the SNPs (rs16891867, rs1861046) are on chromosome 4 (and seem to be linked), as they show the same frequencies in each population listed in the Single Nucleotide Polymorphism Database (dbSNP). The frequency of the risk SNPs are 0.357 in Siberians, 0.200 in Koreans, and 0.042 in northern Swedes. The SNP on chromosome 13 (rs11838918) has an F_{ST} of 0.014, and the risk allele has a frequency ranging from 0.000 to 0.130 in Africa and essentially the same frequency (around 0.030) in the rest of the world. The most strongly differentiated SNP, rs7950811, is on chromosome 11 and has an F_{ST} of 0.135, with a range of 0.294–0.07 in Africa, 0.125–0.00 in Europe, and 0.130–0.00 in East Asia.

59. S. L. Davies, "The Reality of False Confessions—Lessons of the Central Park Jogger Case," *New York University Review of Law and Social Change* 2 (2006): 209–54.

60. Tax Policy Center, Urban Institute and Brookings Institution, Fiscal Facts, "Median Value of Family Net Worth by Race or Ethnicity, 2016," March 11, 2019, https://www.taxpolicycenter.org/fiscal-fact/median-value-wealth-race-ff03112019.

61. B. S. Centerwall, "Race, Socioeconomic Status, and Domestic Homicide," *Journal of the American Medical Association* 273, no. 22 (1995): 1755–58.

62. B. Dong, P. H. Egger, and Y. Guo, "Is Poverty the Mother of Crime? Evidence from Homicide Rates in China," *PLoS One* 15, no. 5 (2020): e0233034, https://journals.plos.org/plosone/article?id=10.1371/journal.pone.0233034.

63. United Nations Office on Drugs and Crime, "Global Study on Homicide, 2019 Edition," https://www.unodc.org/unodc/en/data-and-analysis/global-study-on-homicide.html.

8. DRIVING WHILE BLACK AND OTHER DEADLY REALITIES OF INSTITUTIONAL AND SYSTEMIC RACISM

1. United Church of Christ, "Toxic Wastes and Race at Twenty," accessed May 2, 2021, https://www.ucc.org/environmental-ministries_toxic-waste-20. This is also well explained in the documentary film *Unnatural Causes: Is Inequality Making Us Sick?*,

California Newsreel, accessed May 2, 2021, http://newsreel.org/video/UNNATURAL
-CAUSES.

2. R. L. Wagmiller and R. M. Adelman, "Childhood and Intergenerational Poverty:
The Long-Term Consequences of Growing Up Poor," *Columbia Academic Commons*
(November 2009), https://doi.org/10.7916/D8MP5CoZ.

3. Joe discussed the role of systemic racism in the United States as a factor fuel-
ing the COVID-19 pandemic in J. L. Graves, "Their Money, Our Lives," *Science for
the People*, August 23, 2020, https://magazine.scienceforthepeople.org/web-extras
/covid-19-coronanvirus-wealth-race-public-health/.

4. There is an overabundance of evidence to support this fact. See A. Gelman, J. Fagan,
and A. Kiss, "An Analysis of the New York City Police Department's 'Stop-and-Frisk'
Policy in the Context of Claims of Racial Bias," *Journal of the American Statistical
Association* 102, no. 479 (2007): 813–23, https://doi.org/10.1198/016214506000001040;
J. Rojek, R. Rosenfeld, and S. Decker, "Profiling Race: The Racial Stratification of
Searchers in Police Traffic Stops," *Criminology* 50, no. 4 (2012): 993–1023, https://doi
.org/10.1111/j.1745-9125.2012.00285.x; F. R. Baumgartner, D. A. Epp, and K. Shoub, *Sus-
pect Citizens: What 20 Million Traffic Stops Tell Us About Policing and Race* (Cambridge: Cambridge University Press, 2018).

5. Gelman et al., "An Analysis of the New York City Police Department's 'Stop-and-Frisk'
Policy"; Rojek et al., "Profiling Race"; and Baumgartner et al., *Suspect Citizens.*

6. A classic study of this problem is J. G. Miller, *Search and Destroy: African American Males
in the Criminal Justice System* (Cambridge: Cambridge University Press, 1996). This work
is more than twenty years old but accurately describes much of what is still going on today.

7. Miller, *Search and Destroy.*

8. F. R. Baumgartner, D. A. Epp, and K. Shoub, *Suspect Citizens: What 20 Million Traffic
Stops Tell Us About Policing and Race* (Cambridge: Cambridge University Press, 2018).

9. N. J. Duru, "The Central Park Five, the Scottsboro Boys, and the Myth of the Bestial
Black Man," *Cardozo Law Review* 25 (2004): 1315, available at SSRN: https://papers
.ssrn.com/sol3/papers.cfm?abstract_id=814072.

10. S. L. Davies, "The Reality of False Confessions—Lessons of the Central Park Jogger
Case," *N.Y.U. Review of Law and Social Change* 30, no. 2 (2005–2006): 209.

11. R. Stone, "HHS 'Violence Initiative' Caught in a Crossfire," *Science* 258, no. 5080
(1992): 212–13, https://doi.org/10.1126/science.1411519.

12. B. Rensberger, "Science and Sensitivity," *Washington Post*, March 1, 1992, https://www
.washingtonpost.com/archive/opinions/1992/03/01/science-and-sensitivity/285e7541
-3b66-48c4-9cc9-55fb37d013f9/.

13. J. Silverberg and J. P. Gray, "Violence and Peacefulness as Behavioral Potentialities of
Primates" in *Aggression and Peacefulness in Humans and Other Primates*, ed. J. Silverberg and J. P. Gray (Oxford: Oxford University Press, 1992).

14. P. Dray, *At the Hands of Persons Unknown: The Lynching of Black America* (New York:
Random House, 2002).

15. At the time of this writing, another unarmed black man, Jacob Blake, was shot in the
back seven times in front of his children in Kenosha, Wisconsin; B. Booker, "Man
Shot by Kenosha, Wis., Police Paralyzed from the Waist Down, Lawyer Says," https://
www.npr.org/sections/live-updates-protests-for-racial-justice/2020/08/25/905786759
/another-night-of-clashes-and-unrest-in-kenosha-wis-following-jacob-blake-shootin.

16. D. M. Summerville, *Rape and Race in the Nineteenth-Century South* (Chapel
Hill: University of North Carolina Press, 2004); and M. Marable, *How Capitalism*

Underdeveloped Black America: Problems in Race, Political Economy, and Society (Chicago: Haymarket, 2015).

17. P. Wagner and W. Sawyer, "States of Incarceration: The Global Context 2018," Prison Policy Initiative, June 2018, https://www.prisonpolicy.org/global/2018.html.

18. T. Mendelberg, "Executing Hortons: Racial Crime in the 1988 Presidential Campaign," *Public Opinion Quarterly* 61, no. 1 (1997): 134–57, https://doi.org/10.1086/297790.

19. J. Esherick, *Prison Rehabilitation: Success Stories and Failures* (Broomall, PA: Mason Crest, 2015).

20. M. Brickner and S. Diaz, "Prisons for Profit: Incarceration for Sale," 38 Human Rights Magazine 13 (July 1, 2011).

21. D. Pager, *Marked: Race, Crime, and Finding Work in the Era of Mass Incarceration* (Chicago: University of Chicago Press, 2007).

22. J. Tucker, "Captive Lives," *San Francisco Chronicle*, September 16, 2016, https://projects.sfchronicle.com/2016/captive-lives/.

23. A. Nellis, "The Color of Justice: Racial and Ethnic Disparity in State Prisons," The Sentencing Project, June 14, 2016, https://www.sentencingproject.org/publications/color-of-justice-racial-and-ethnic-disparity-in-state-prisons/.

24. M. Alexander, *The New Jim Crow: Mass Incarceration in the Age of Colorblindness* (New York: New Press, 2010); B. Stevenson, *Just Mercy: A Story of Justice and Redemption* (New York: One World, 2014).

25. R. Balko, *Rise of the Warrior Cop: The Militarization of America's Police Forces* (New York: Public Affairs, 2013).

26. National Center for Education Statistics, Status and Trends in the Education of Racial and Ethnic Groups, "Indicator 27: Educational Attainment," updated February 2019, https://nces.ed.gov/programs/raceindicators/indicator_RFA.asp.

27. H. Lamb, *The Crusades* (New York: Bantam, 1967, originally published in 1931).

28. A. Jensen, "How Much Can We Boost IQ and Scholastic Achievement?," *Harvard Educational Review* 39, no. 1 (1969): 1–123.

29. J. Kozol, *Savage Inequalities: Children in America's Schools* (New York: Broadway Books, 2012).

30. S. Morton, *Crania Americana; or, a Comparative View of Skulls of Various Aboriginal Nations of North and South America; to which Is Prefixed an Essay on the Varieties of the Human Species* (Philadelphia: J. Dobson, 1839); and R. Herrnstein and C. R. Murray, *The Bell Curve: Intelligence and Class Structure in American Life* (New York: Free Press, 1994).

31. M. Bertrand and S. Mullainthan, "Are Emily and Greg More Employable Than Lakisha and Jamal? A Field Experiment on Labor Market Discrimination," *American Economic Review* 94, no. 4 (2004): 991–1013.

32. J. L. Graves and E. Jarvis, "An Open Letter: Scientists and Racial Justice," *Scientist*, June 19, 2020, https://www.the-scientist.com/news-opinion/an-open-letter-scientists-and-racial-justice-67648.

33. H. Queneau and A. Sen, "On the Structure of US Unemployment Disaggregated by Race, Ethnicity, and Gender," *Economics Letters* 117, no. 1 (October 1, 2012): 91–95, https://doi.org/10.1016/j.econlet.2012.04.065.

34. Centers for Disease Control and Prevention, "Morbidity and Mortality Weekly Reports," 54, no. 20 (May 27, 2005): 513–516.

35. United Church of Christ, "Toxic Wastes and Race at Twenty."

36. B. Chappell and S. Nuyen, "2 Hurricanes Could Form in the Gulf of Mexico Next Week— An Apparent First," NPR, https://www.npr.org/2020/08/21/904674353/2-hurricanes-could-be-in-gulf-coast-early-next-week-weather-service-says.

37. ESRI, "Aftermath of Katrina: A Time of Environmental Racism," accessed May 2, 2021, https://www.arcgis.com/apps/Cascade/index.html?appid=2106693b39454f0eb0abc5c2 ddf9ce40.

38. J. Hanna and S. Chan, "California Wildfires Have Burned 1.25 Million Acres, But Fire-fighters Say the Weather Is Now Helping," CNN, August 25, 2020, https://www.cnn .com/2020/08/25/us/california-fires-tuesday/index.html.

39. J. E. Keely, H. Safford, C. J. Fotheringham, J. Franklin, and M. Moritz, "The 2007 Southern California Wildfires: Lessons in Complexity," *Journal of Forestry* 107, no. 6 (2009): 287–96, https://doi.org/10.1093/jof/107.6.287.

40. J. L. Graves, *The Race Myth: Why We Pretend Race Exists in America* (New York: Dutton, 2004), 194–203.

41. K. McIntosh, E. Moss, K. Moss, R. Nunn, J. Shambaugh, "Up Front: Examining the Black-White Wealth Gap," Brookings, February 27, 2020, https://memphis.uli.org/wp -content/uploads/sites/49/2020/07/Examining-the-Black-white-wealth-gap.pdf.

42. N. T. Sharma, *Hip Hop Desis: South Asians, Blackness, and a Global Race Consciousness* (Durham, NC: Duke University Press, 2010).

43. L. Salyer, *Laws as Harsh as Tigers: Chinese Immigrants and the Shaping of Modern Immigration Law* (Chapel Hill: University of North Carolina Press, 1995).

9. DNA AND ANCESTRY TESTING

1. A. Regalado, "More Than 26 Million People Have Taken an At-Home Ancestry Test," *MIT Technology Review*, February 11, 2019, https://www.technologyreview .com/2019/02/11/103446/more-than-26-million-people-have-taken-an-at-home -ancestry-test/.

2. C. M. Rands, S. Meader, C. P. Ponting, and G. Lunter, "8.2 Percent of the Human Genome is Constrained: Variation in Rates of Turnover Across Functional Element Classes in the Human Lineage," *PLoS Genetics* 10, no. 7 (2014): e1004525, https://doi .org/10.1371/journal.pgen.1004525.

3. G. Barbujani and C. Colonna, "Human Genome Diversity: Frequently Asked Questions," *Trends in Genetics* 26, no. 7 (2010): 285–95, https://doi.org/10.1016/j.tig .2010.04.002.

4. K. Bryc, E. Y. Durand, J. M. Macpherson, D. Reich, and J. L. Mountain, "The Genetic Ancestry of African Americans, Latinos, and European Americans Across the United States," *American Journal of Human Genetics* 96, no. 1 (2015): 37–53, https://doi .org/10.1016/j.ajhg.2014.11.010; and S. J. Micheletti, K. Bryc, S. G. Ancona Esselmann, et al., "Genetic Consequences of the Transatlantic Slave Trade in the Americas," *American Journal of Human Genetics* 107, no. 2 (2020): 265–77, https://doi.org/10.1016/j .ajhg.2020.06.012. The first study utilized the V1, V2 variants of the Illumina Human-Hap 550 + Bead Chip, which included 25,000 custom SNPs chosen by 23andMe, total-ing 560,000 SNPs. The researchers also used the V3 platform Illumina OmniExpress + Bead Chip + custom content, which included 950,000 SNPs; and the V4 platform, which had an additional custom 510,000 SNPs.

5. You can watch world population grow in real time at the Current World Population Clock, https://www.worldometers.info/world-population/.

6. E. Jehaes, H. Pfeiffer, K. Toprak, et al., "Mitochondrial DNA Analysis of the Putative Heart of Louis XVII, Son of Louis XVI and Marie-Antoinette," *European Journal of Human Genetics* 9, no. 3 (2001): 185–90, https://doi.org/10.1038/sj.ejhg.5200602.

7. A. Gordon Reed, *Thomas Jefferson and Sally Hemings: An American Controversy* (Charlottesville: University Press of Virginia, 1998).

8. E. A. Foster, M. A. Jobling, P. G. Taylor, et al., "Jefferson Fathered Slave's Last Child," *Nature* 396, no. 6706 (1998): 27–28, https://doi.org/10.1038/23835.

9. S. A. Tishkoff, F. A. Reed, F. R. Friedlaender, et al., "The Genetic Structure and History of Africans and African Americans," *Science* 324, no. 5930 (2009): 1035–44, https://doi.org/10.1126/science.1172257.

10. B. M. Henn, C. R. Gignoux, M. Jobin, et al., "Hunter-Gatherer Genomic Diversity Suggests a Southern African Origin for Modern Humans," *Proceedings of the National Academy of Sciences* 108, no. 13 (2011): 5154–62, https://doi.org/10.1073/pnas.1017511108.

11. P. Manning, "Slavery and the Slave Trade in West Africa: 1450–1930," in *The History of African Development. An Online Textbook for a New Generation of African Students and Teachers*, ed. E. Frankema, E. Hillbom, U. Kufakurinani, and F. Meier zu Selhausen, African Economic History Network, accessed May 4, 2021, https://www.aehnetwork.org/textbook/; also see P. Lovejoy, "The Impact of the Atlantic Slave Trade on Africa: A Review of the Literature," *Journal of African History* 30, no. 3 (1989): 365–94, https://doi.org/10.1017/S0021853700024439.

12. A. M. Huml, C. Sullivan, M. Figueroa, K. Scott, and A. R. Sehgal, "Consistency of Direct-to-Consumer Genetic Testing Results Among Identical Twins," *American Journal of Medicine* 133, no. 1 (2020): 143–46.e2, https://doi.org/10.1016/j.amjmed.2019.04.052.

13. J. M. Lind, H. B. Hutcheson-Dilks, S. M. Williams, et al., "Elevated Male European and Female African Contributions to the Genomes of African American Individuals," *Human Genetics* 120, no. 5 (2007): 713–22, https://doi.org/https://doi.org/10.1007/s00439-006-0261-7.

14. Micheletti et al., "Genetic Consequences of the Transatlantic Slave Trade in the Americas."

15. L. M. Holson, "Ancestry.com Apologizes for Ad Showing Slavery-Era Interracial Couple," *New York Times*, April 19, 2019, https://www.nytimes.com/2019/04/19/us/ancestry-dna-slavery-commercial.html.

16. The story of Patsy is well recounted in Solomon Northup's *Twelve Years a Slave* (New York: Graymalkin Media, 2014), originally published in 1853. This character was also brought to the screen in the 2013 film *Twelve Years a Slave* with Kenyan-born actress Lupita Nyong'o in the role of Patsy. (By the way, Kenyans were never sold in the transatlantic slave trade.)

17. V. E. Bynum, "'White Negroes' in Segregated Mississippi: Miscegenation, Racial Identity, and the Law," *Journal of Southern History* 64, no. 2 (1998): 247–76, https://doi.org/10.2307/2587946; also see R. Grant, "The True Story of 'Free State of Jones,'" *Smithsonian* (March 2016), https://www.smithsonianmag.com/history/true-story-free-state-jones-180958111/.

18. A. Panofsky and J. Donovan, "Genetic Ancestry Testing Among White Nationalists: From Identity Repair to Citizen Science," *Social Studies of Science* 49, no. 5 (2019): 653–81, https://doi.org/10.1177/0306312719861434.

19. M. Slatkin and F. Racimo, "Ancient DNA and Human History," *Proceedings of the National Academy of Sciences* 113, no. 23 (2016): 6380–87, https://doi.org/10.1073/pnas.1524306113.

20. L. Cavalli-Sfroza, P. Menozzi, and A. Piazza, *The History and Geography of Human Genes* (Princeton, NJ: Princeton University Press, 1994).

21. D. Reich, *Who We Are and How We Got Here: Ancient DNA and the New Science of the Human Past* (New York: Pantheon, 2018).

22. Reich, *Who We Are and How We Got Here*.

23. S. Sankararaman, S. Mallick, N. Patterson, and D. Reich, "The Combined Landscape of Denisovan and Neanderthal Ancestry in Present-Day Humans," *Current Biology* 26, no. 9 (2016): 1241–47, https://doi.org/10.1016/j.cub.2016.03.037.

24. E. Huerta-Sánchez, X. Jin, Asan, et al., "Altitude Adaptation in Tibetans Caused by Introgression of Denisovan-Like DNA," *Nature* 512, no. 7513 (2014): 194–97, https://doi.org/10.1038/nature13408.

25. Huerta-Sánchez et al., "Altitude Adaptation in Tibetans."

26. E. M. Oziolor, N. M. Reid, S. Yair, et al., "Adaptive Introgression Enables Evolutionary Rescue from Extreme Environmental Pollution," *Science* 364, no. 6439 (2019): 455–57, https://doi.org/10.1126/science.aav4155.

27. B. Kislev, "Neanderthal and Woolly Mammoth Molecular Resemblance: Genetic Similarities May Underlie Cold Adaptation Suite," *Human Biology* 90, no. 2 (2019): 115–28, https://doi.org/10.13110/humanbiology.90.2.03.

28. J. K. Pickrell and D. Reich, "Toward a New History and Geography of Human Genes Informed by Human DNA," *Trends in Genetics* 30, no. 9 (2014): 377–89, https://doi.org/10.1016/j.tig.2014.07.007.

29. R. Nielsen, J. M. Akey, M. Jakobsson, et al., "Tracing the Peopling of the World Through Genomics," *Nature* 541, no. 7637 (2017): 302–10, https://doi.org/10.1038/nature21347.

30. M. F. Hammer, A. E. Woerner, F. L. Mendez, J. C. Watkins, and J. D. Wall, "Genetic Evidence for Archaic Admixture in Africa," *Proceedings of the National Academy of Sciences* 108, no. 37 (2011): 15123–128, https://doi.org/10.1073/pnas.1109300108.

31. T. Günther and M. Jakobsson, "Genes Mirror Migrations and Cultures in Prehistoric Europe—A Population Genomic Perspective," *Current Opinion in Genetics & Development* 41 (2016): 115–23, https://doi.org/10.1016/j.gde.2016.09.004.

32. T. Günther, C. Valdiosera, H. Malmström, et al., "Ancient Genomes Link Early Farmers from Atapuerca in Spain to Modern-Day Basques," *Proceedings of the National Academy of Sciences* 112, no. 38 (2015): 11917–922, https://doi.org/10.1073/pnas.1509851112.

33. E. R. Jones, G. Gonzalez-Fortes, S. Connell, et al., "Upper Palaeolithic Genomes Reveal Deep Roots of Modern Eurasians," *Nature Communications* 6 (2015): 8912, https://doi.org/10.1038/ncomms9912.

34. M. Rasmussen, X. Guo, Y. Wang , et al., "An Aboriginal Australian Genome Reveals Separate Human Dispersals Into Asia," *Science* 334, no. 6052 (2011): 94–98, https://doi.org/10.1126/science.1211177.

35. A. S. Malaspinas, M. C. Westaway, C. Muller, et al., "A Genomic History of Aboriginal Australia," *Nature* 538, no. 7624 (2016): 207–14, https://doi.org/10.1038/nature18299.

36. M. Raghavan, M. Steinrücken, K. Harris, et al., "Genomic Evidence for the Pleistocene and Recent Population History of Native Americans," *Science* 349, no. 6250 (2015): aab3884, https://doi.org/10.1126/science.aab3884.

37. K. L. Hunley and G. S. Cabana, "Beyond Serial Founder Effects: The Impact of Admixture and Localized Gene Flow on Patterns of Regional Genetic Diversity," *Human Biology* 88, no. 3 (2016): 219–31, https://doi.org/10.13110/humanbiology.88.3.0219.

38. R. Kuper and S. Kropelin, "Climate-Controlled Holocene Occupation in the Sahara: Motor of Africa's Evolution," *Science* 313, no. 5788 (2006): 803–7, https://doi.org/10.1126/science.1130989.

39. S. Fan, M. E. Hansen, Y. Lo, and S. A. Tishkoff, "Going Global by Adapting Local: A Review of Recent Human Adaptation," *Science* 354, no. 6308 (2016): 54–59, https://doi .org/10.1126/science.aaf5098.

40. D. Smay and G. Armelagos, "Galileo Wept: A Critical Assessment of the Use of Race in Forensic Anthropology," *Transforming Anthropology* 9, no. 2 (2000): 19–29, https:// doi.org/10.1525/tran.2000.9.2.19.

41. E. Giles and O. Elliot, "Race Identification from Cranial Measurements," *Journal of Forensic Sciences* 7, no. 2 (1962): 147–57.

42. A. H. Goodman, "Bred in the Bone?," *The Sciences* 37 (1997): 20–25, https://doi .org/10.1002/j.2326-1951.1997.tb03296.x.

43. R. Lewis, *Human Genetics: Concepts and Applications*, 5th ed. (Boston: McGraw-Hill, 2003).

44. C. Phillips, "The Golden State Killer Investigation and the Nascent Field of Forensic Genealogy," *Forensic Science International: Genetics* 36 (2018): 186–88, https://doi .org/10.1016/j.fsigen.2018.07.010.

45. B. Scheck, "The Innocence Project at Twenty: An Interview with Barry Scheck. Interview by Jane Gitschier," *PLoS Genetics* 9, no. 8 (2013): e1003692, https://doi.org/10.1371 /journal.pgen.1003692.

46. P. A. Chow-White and T. Duster, "Do Health and Forensic DNA Databases Increase Racial Disparities?," *PLoS Medicine* 8, no. 10 (2011): e1001100, https://doi.org/10.1371 /journal.pmed.1001100.

47. A. O. Amankwaa and C. McCartney, "The UK National DNA Database: Implementation of the Protection of Freedoms Act 2012," *Forensic Science International* 284 (2018): 117–28, https://doi.org/10.1016/j.forsciint.2017.12.041.

48. S.-L. Wee, "China Is Collecting DNA from Tens of Millions of Men and Boys, Using U.S. Gear," *New York Times*, June 17, 2020, https://www.nytimes.com/2020/06/17/world /asia/China-DNA-surveillance.html.

10. RACE NAMES AND "RACE MIXING"

1. J. M. Fish, "Mixed Blood: An Analytical Look at Methods of Classifying Race," *Psychology Today*, November 1, 1995, https://www.psychologytoday.com/us/articles/199511 /mixed-blood.

2. D. I. Ketzer and D. Arel, "Censuses, Identity Formation, and the Struggle for Political Power," in *Census and Identity: The Politics of Race, Ethnicity, and Language in National Census*, ed. D. I. Ketzer and D. Arel (New York: Cambridge University Press, 2002).

3. See Office of Management and Budget Directive 15, https://wonder.cdc.gov/WONDER /help/populations/bridged-race/directive15.html.

4. S. Arasaratnam, "Weavers, Merchants and Company: The Handloom Industry in Southeastern India 1750–1790," *Indian Economic and Social History Review* 12, no. 3 (1980): 257–81.

5. F. V. Harrison, "Unraveling Race for the Twenty-First Century," in *Exotic No More: Anthropology for the Contemporary World*, 2nd ed., ed. J. MacClancy (Chicago: University of Chicago Press, 2019).

6. C. C. Gravlee and W. W. Dressler, "Emic Ethnic Classification in Southeastern Puerto Rico: Cultural Consensus and Semantic Structure," *Social Forces* 83 (2005): 949–70;

R. L. Reichmann, *Race in Contemporary Brazil: From Indifference to Inequality* (State College, PA: Pennsylvania State University Press, 2010).

7. L. Wright Jr., "Who's Black, Who's White, and Who Cares: Reconceptualizing the United States's Definition of Race and Racial Classifications," *Vanderbilt Law Review* 48, no. 2 (1995): 513.

8. T. Pegelow, "Determining 'People of German Blood', 'Jews' and 'Mischlinge': The Reich Kinship Office and the Competing Discourses and Powers of Nazism, 1941–1943," *Contemporary European History* 15, no. 1 (2006): 43–65, https://doi.org/10.1017/S0960777306003092.

9. K. Bryc, E. Y. Durand, J. M. Macpherson, D. Reich, and J. L. Mountain, "The Genetic Ancestry of African Americans, Latinos, and European Americans Across the United States," *American Journal of Human Genetics* 96, no. 1 (2015): 37–53, https://doi.org/10.1016/j.ajhg.2014.11.010.

10. J. Brown (with P. Ellis), *Say It Loud—I'm Black and I'm Proud*, https://en.wikipedia.org/wiki/Say_It_Loud_%E2%80%93_I%27m_Black_and_I%27m_Proud.

11. Bryc et al., "The Genetic Ancestry of African Americans, Latinos, and European Americans."

12. L. R. Arana, "The Exploration of Florida and Sources on the Founding of St. Augustine," *Florida Historical Quarterly* 44, no. 1/2 (1965): 1–16; R. J. Weber, *Myth and History of the Hispanic Southwest* (Albuquerque: University of New Mexico Press, 1988); and J. L. Allen, "From Cabot to Cartier: The Early Exploration of Eastern North America, 1497–1543," *Annals of the Association of American Geographers* 82, no. 3 (1992): 500–21, https://doi.org/10.1111/j.1467-8306.1992.tb01972.x.

13. J. W. Davidson, B. DeLay, C. L. Heyman, M. H. Lytle, and M. B. Stoff, *Nation of Nations: A Narrative History of the American Republic* (Boston: McGraw-Hill Higher Education, 2008).

14. S. Kelley, "'Mexico in His Head': Slavery and the Texas-Mexico Border, 1810–1860," *Journal of Social History* 37, no. 3 (2004): 709–23, https://doi.org/10.1353/jsh.2004.0010; and J. D. P. Fuller, "The Slavery Question and the Movement to Acquire Mexico, 1846–1848," *Mississippi Valley Historical Review* 21, no. 1 (1934): 31–48.

15. E. Guzman, "*Igartua de la Rosa v. United States*: The Right of the United States Citizens of Puerto Rico to Vote for the President and the Need to Re-Evaluate America's Territorial Policy," *Journal of Constitutional Law* 14, no. 1 (2001), https://scholarship.law.upenn.edu/jcl/vol4/iss1/3.

16. P. Cabán, *Constructing a Colonial People: Puerto Rico and the United States: 1898–1932*,(New York: Routledge, 1999).

17. "Census Bureau Releases Updates on Race Categories for 2020 Census," https://2020census.gov/en/about-questions/2020-census-questions-race.html.

18. These race categories are found at U.S. Bureau of the Census, https://2020census.gov/en/about-questions/2020-census-questions-race.html.

19. W. D. Roth, Ş. Yaylacı, K. Jaffe, and L. Richardson, "Do Genetic Ancestry Tests Increase Racial Essentialism? Findings from a Randomized Controlled Trial," *PLoS ONE* 15, no. 1 (2020): e0227399, https://doi.org/10.1371/journal.pone.0227399.

20. M. Omi and H. Winant, "Racial Formations," in *Race, Class, and Gender in the United States*, ed. P. S. Rothenberg (New York: Worth, 2007).

21. S. Hoffman, "Is There a Place for Race as a Legal Concept," *Faculty Publications* 227 (2004), https://scholarlycommons.law.case.edu/faculty_publications/227.

22. W. Booth, "Phobia About Blacks Brings Worker's Compensation Award," *Washington Post*, August 13, 1992, https://www.washingtonpost.com/archive/politics/1992/08/13

/phobia-about-blacks-brings-workers-compensation-award/5bff7057-22c3-4b45
-ad32-0f58d0fd9b1f/.

23. P. Buescher, Z. Gizlice, and K. Jones-Vessey, "Self-Reported Versus Published Data on Racial Classification in North Carolina Birth Records," *SCHS Studies* no. 139—Racial Classification (February 2004), https://schs.dph.ncdhhs.gov/schs/pdf/schs139.pdf.

24. R. A. Hahn, "Why Race Is Differentially Classified on U.S. Birth and Infant Death Certificates: An Examination of Two Hypotheses," *Epidemiology* 10, no. 2 (1999): 108–11.

25. E. Hackenmueller, "The (Mis)Representation of Interracial Couples in Television Advertisements" (master's thesis, University of Alabama, Department of Advertising and Public Relations, 2020), http://ir.ua.edu/handle/123456789/6982.

26. K. Parker, J. M. Horowitz, R. Morin, and M. H. Lopez, "Multiracial in America: Proud, Diverse, and Growing in Numbers," Pew Research Center, June 11, 2015, https://www.pewsocialtrends.org/2015/06/11/multiracial-in-america/.

27. J. M. Chen, N. S. Kteily, and A. K. Ho, "Whose Side Are You On? Asian American Distrust of Asian-White Biracials Predicts More Exclusion from the Ingroup," *Personality and Social Psychology Bulletin* 45, no. 6 (2019): 827–41, https://doi.org/10.1177/0146167218798032.

28. S. Outram, J. L. Graves, J. Powell, et al., "Genes, Race, and Causation: US Public Perspectives About Racial Difference," *Race Soc. Probl.* (2018), https://doi.org/10.1007/s12552-018-9223-7.

29. A discussion of hybrid vigor or heterosis can be found in any standard genetics text; e.g., see D. S. Falconer and T. Mackay, *Introduction to Quantitative Genetics*, 4th ed. (Essex, UK: Addison Wesley Longman, 1996), 258.

30. C. Darwin, *The Descent of Man and Selection in Relation to Sex* (Princeton, NJ: Princeton University Press, 1981), first published in 1871. His discussion of the fertility of the hybrids is on 221–23.

31. K. Wailoo, *Sickle Cell Disease: Dying in the City of the Blues* (Chapel Hill: University of North Carolina Press, 2001); and J. H. Jones, *Bad Blood: The Tuskegee Syphilis Experiment*, new and exp. ed. (New York: Free Press, 1993).

32. C. B. Davenport and M. Steggerda, *Race Crossing in Jamaica* (Washington, DC: Carnegie Institution of Washington, 1929).

33. See discussion of Charles Davenport and the Eugenics Record Office in J. L. Graves, *The Emperor's New Clothes: Biological Theories of Race at the Millennium* (New Brunswick, NJ: Rutgers University Press, 2005), 115–25.

34. A discussion of how this sterilization was conducted can be found in S. F. Weiss, "The Race Hygiene Movement in Germany 1904–1945," in *The Wellborn Science: Eugenics in Germany, France, Brazil, and Russia*, ed. M. Adams (New York: Oxford University Press, 1990).

35. WedMd, "Multiple Myeloma: Causes and Risk Factors," https://www.webmd.com/cancer/multiple-myeloma/are-you-at-risk-for-multiple-myeloma.

36. S. A. Tishkoff, F. A. Reed, F. R. Friedlaender, et al., "The Genetic Structure and History of Africans and African Americans," *Science* 324, no. 5930 (2009): 1035–44, https://doi.org/10.1126/science.1172257.

37. J. Nott and G. R. Gliddon, *Indigenous Races of the Earth or New Chapters of Ethnological Inquiry* (Philadelphia: Lippincott and Grambo, 1857).

38. J. Snow, "The Civilization of White Men: The Race of the Hindu in *United States v. Bhagat Singh Thind*," in *Race, Nation, and Religion in the Americas*, ed. H. Goldschmidt and E. McAllister (New York: Oxford University Press, 2004).

39. J. C. Prichard, *Researches Into the Physical History of Mankind, vol. III. Researches Into the Origin of the European Nations* (London: Sherwood, Gilbert, and Piper, 1841).

40. R. Dart, "*Australopithecus africanus*: The Man-Ape of South Africa," *Nature* 115, no. 2884 (1925).

41. J. K. McKee, F. E. Poirier, and W. S. McGraw, *Understanding Human Evolution*, 5th ed. (New York: Routledge, 2004).

42. C. Coon, *The Origin of Races* (New York: Knopf, 1962).

43. A. Montagu, *Mankind's Most Dangerous Myth: The Fallacy of Race*, 6th ed. (Palo Alto, CA: Alta Mira, 1997). He quotes an interview by Herman Rauschning with Adolf Hitler; see H. Rauschning, *The Voice of Destruction* (New York: Putnam, 1940).

44. M. Kaback, J. Lim-Steele, D. Dabholkar, et al., "Tay-Sachs Disease—Carrier Screening, Prenatal Diagnosis, and the Molecular Era. An International Perspective, 1970 to 1993. The International TSD Data Collection Network," *Journal of the American Medical Association* 270, no. 19 (1990): 2307–15.

45. Montagu, *Mankind's Most Dangerous Myth*.

46. P. Kolchin, "Whiteness Studies: The New History of Race in America," Journal of American History 89, no. 1 (2002): 154–73, https://doi.org/10.2307/2700788.

47. See the classic historical work on this phenomenon, D. R. Roediger, *Working Toward Whiteness: How America's Immigrants Became White; The Strange Journey from Ellis Island to the Suburbs* (New York, Basic Books, 2005).

48. Joe is not alone in thinking this way; see J. Barndt, *Becoming an Anti-Racist Church: Journeying Towards Wholeness* (Minneapolis: Fortress, 2011).

49. P. Bump, "Presenting the Least Misleading Map of the 2016 Election," *Washington Post*, July 30, 2018, https://www.washingtonpost.com/news/politics/wp/2018/07/30/presenting-the-least-misleading-map-of-the-2016-election/.

50. T. T. Reny, L. Collingwood, and A. A. Valenzuela, "Vote Switching in the 2016 Election: How Racial and Immigration Attitudes, Not Economics, Explain Shifts in White Voting," *Public Opinion Quarterly* 83, no. 1 (2019): 91–113, https://doi.org/10.1093/poq/nfz011.

51. J. Daniels, *Cyber Racism: White Supremacy Online and the New Attack on Civil Rights* (Lanham, MD: Rowman & Littlefield, 2009).

52. "Fraternity Activities Suspended at Syracuse University Following Latest Racist Incident, Investigation Into Graffiti Ongoing," DiversityInc, November 18, 2019, https://www.diversityinc.com/syracuse-university-racist-graffiti/?utm_source=hs_email&utm_medium=email&utm_content=79571173&_hsenc=p2ANq%E2%80%80%A6.

53. R. Ralston, M. Motta, and J. Spindel, "When OK Is Not OK: Public Concern About White Nationalism in the U.S. Military," *Armed Forces & Society* (April 2020): 1–12, https://doi.org/10.1177/0095327X20918394.

54. Daniels, *Cyber Racism*.

55. J. D. Mayer, *Running on Race: Racial Politics in the Presidential Campaigns, 1960–2000* (New York: Random House, 2002). The Nixon quote is on 69.

56. P. Sullivan, *Days of Hope: Race and Democracy in the New Deal Era* (Chapel Hill: University of North Carolina Press, 1996).

57. Southern Poverty Law Center, "White Nationalist," https://www.splcenter.org/fighting-hate/extremist-files/ideology/white-nationalist.

58. E. Levenson, "The Realities of Being a Black Bird Watcher," CNN, May 27, 2020, https://www.cnn.com/2020/05/27/us/birdwatching-black-christian-cooper/index.html.

59. T. McCoy, "In Jim Cooley's Open-Carry America, Even a Trip to Walmart Can Require an AR-15," *Washington Post*, September 17, 2016, https://www.washingtonpost.com/national/guns-and-sodas/2016/09/17/805e0db4-79e9-11e6-bd86-b7bbd53d2b5d_story.html.

60. M. Macaya, M. Wagner, and M. Hayes, "The Latest on Kenosha Police Shooting of Jacob Blake," CNN, August 31, 2020, https://www.cnn.com/us/live-news/jacob-blake-kenosha-police-shooting-08-31-2020/index.html.

61. M. Holcombe, S. Nottingham, and E. Levenson, "Man Killed in Portland Shooting Identified by Police as Aaron J. Danielson," CNN, August 31, 2020, https://www.cnn.com/2020/08/31/us/governor-brown-portland-plan-curb-protests/index.html.

62. A. Wise, "Trump Defends Kenosha Shooting Suspect," NPR, August 31, 2020, https://www.npr.org/sections/live-updates-protests-for-racial-justice/2020/08/31/908137377/trump-defends-kenosha-shooting-suspect.

11. A WORLD WITHOUT RACISM?

1. G. Childress, "State Board of Education OKs New Social Study Standards; Lt. Governor Calls It 'Irresponsible,'" NC Policy Watch, February 5, 2021, http://pulse.ncpolicywatch.org/2021/02/05/state-board-of-education-oks-new-social-study-standards-lt-governor-calls-it-irresponsible/#sthash.jKcKVCHx.dpbs.

2. W. Guzman, "Michigan State Vice President of Research Stephen Hsu Resigns," State News, June 19, 2020, https://statenews.com/article/2020/06/michigan-state-vp-of-research-stephen-hsu-resigns.

3. A. F. Poiussaint, "Is Extreme Racism a Mental Disorder?," *Western Journal of Medicine* 176, no. 4 (2002).

4. S. Barnes, Washington Staff, and Associated Press, "Commitment March on Washington: Crowds at National Mall Call for Justice," NBC, August 28, 2020, https://www.nbcwashington.com/news/commitment-march-on-washington-2020-thousands-to-converge-on-national-mall/2403921/.

5. "Trump Holds NH Rally After Accepting Republican Presidential Nomination," NBC, August 28, 2020, https://www.nbcboston.com/news/politics/decision-2020/trump-to-hold-rally-in-nh-friday/2185378/.

6. As of January 8, 2021.

7. One of our favorites is T. Schick Jr., and L. Vaughan, *How to Think About Weird Things: Critical Thinking for a New Age*, 5th ed. (Boston: McGraw-Hill, 2008).

8. L. M. Bartels, "Ethnic Antagonism Erodes Republicans' Commitment to Democracy," *Proceedings of the National Academy of Sciences* 117, no. 37 (2020): 22752–759, https://doi.org/10.1073/pnas.2007747117.

9. D. Searcey, "The Battle Over Biscuits and Gravy at the 11-Worth Cafe," *New York Times*, September 5, 2020, https://www.nytimes.com/2020/09/05/us/politics/omaha-cafe-confederate-protests.html?searchResultPosition=1.

10. B. F. Schaffner, M. Macwilliams, and T. Nteta, "Understanding White Polarization in the 2016 Vote for President: The Sobering Role of Racism and Sexism," *Political Science Quarterly* 133, no. 1 (2018): 9–34, https://doi.org/10.1002/polq.12737.

11. S. Coll, "The Case for Dumping the Electoral College," *New Yorker*, September 13, 2020, https://www.newyorker.com/magazine/2020/09/21/the-case-for-dumping-the-electoral-college.

12. Joe once participated in such training for the Portland Oregon police force (commanders and sergeants).

13. See the American Civil Liberties Union (ACLU) bail reform site, https://www.aclu.org/issues/smart-justice/bail-reform.

14. B. Stevenson, *Just Mercy: A Story of Justice and Redemption* (New York: Random House, 2014).

15. J. L. Eberhardt, P. G. Davies, V. J. Purdie-Vaughns, and S. L. Johnson, "Looking Deathworthy: Perceived Stereotypicality of Black Defendants Predicts Capital-Sentencing Outcomes," *Cornell Law Faculty Publications* 41 (2006), https://scholarship.law.cornell.edu/lsrp_papers/41.

16. Hamilton Project, "Rate of Drug Use and Sales, by Race; Rates of Drug Related Criminal Justice Measures, by Race," October 26, 2016, https://www.hamiltonproject.org/charts/rates_of_drug_use_and_sales_by_race_rates_of_drug_related_criminal_justice.

17. J. Hudak, "Colorado's Rollout of Legal Marijuana Is Succeeding: A Report on the State's Implementation of Legalization," *Case Western Reserve Law Review* 65, no. 3 (2015): 649.

18. A. Selsky, "Oregon 1st State to Decriminalize Possession of Hard Drugs," *Associated Press*, February 1, 2021, https://www.usnews.com/news/politics/articles/2021-02-01/oregon-1st-state-to-decriminalize-possession-of-hard-drugs.

19. M. L. King Jr., "Our God Is Marching On" (speech), American Public Media, March 25, 1965, http://americanradioworks.publicradio.org/features/prestapes/mlk_speech.html.

20. W. C. Whatley, "African-American Strikebreaking from the Civil War to the New Deal," *Social Science History* 17, no. 4 (1993): 525–58.

21. C. J. Walley, "Trump's Election and the 'White Working Class': What We Missed," *American Ethnologist* 44, no. 2 (2017): 231–36, https://doi.org/10.1111/amet.12473.

22. K. Harris, A. Kimson, and A. Schwedel, "Labor 2030: The Collision of Demographics, Automation and Inequality," Bain & Company, February 7, 2018, http://www.bain.com/publications/articles/labor-2030-the-collision-of-demographics-automation-and-inequality.aspx.

23. A. L. Kalleberg, "Precarious Work, Insecure Workers: Employment Relations in Transition," *American Sociological Review* 74 (2009): 1–22.

24. J. Prassl, *Humans as a Service: The Promise and Peril of Work in the Gig Economy* (New York: Oxford University Press, 2018).

25. E. P. Bettinger and B. T. Long, "Does Cheaper Mean Better? The Impact of Using Adjunct Instructors on Student Learning Outcomes," *Review of Economics and Statistics* 92, no. 3 (2010): 598–613.

26. R. Wallace, *Dead Epidemiologists: On the Origins of COVID-19* (New York: Monthly Review Press, 2021).

27. B. Donovan, "Reclaiming Race as a Topic of the U.S. Biology Textbook Curriculum," *Science Education* 99, no. 6 (2015): 1092–117.

28. R. Bifulco and H. F. Ladd, "School Choice, Racial Segregation, and Test-Score Gaps: Evidence from North Carolina's Charter School Program," *Journal of Policy Analysis and Management* 26, no. 1 (2006): 31–56.

29. S. Neiman, *Learning from the Germans* (New York: Farrar, Straus and Giroux, 2019).

30. P. Hayner, Unspeakable Truths: Transitional Justice and the Challenge of Truth Commissions (New York: Routledge, 2011).

31. See news coverage, including J. Zelizer, "Why President Trump Is Targeting the 1619 Project," CNN, September 25, 2020, https://www.cnn.com/2020/09/21/opinions/patriotic-education-1776-commission-zelizer/index.html.

32. B. Smedley, A. Y. Stith, and A. R. Nelson, eds., *Unequal Treatment: Confronting Racial and Ethnic Disparities in Health Care* (Washington, DC: National Academies Press, 2004); and P. R. Rose, *Health Equity, Diversity, and Inclusion: Context, Controversies, and Solutions* (Burlington, MA: Jones & Bartlett Learning, 2021).

33. See J. Fodeman and P. Factor, "Solutions to the Primary Care Physician Shortage," *American Journal of Medicine* 128, no. 8 (2015): 800–801, doi:10.1016/j.amjmed.2015.02.023; R. B. Cooper, "What Does It Mean to Have a Physician Shortage?," *Journal of the American Academy of Physician Assistants* 28, no.3 (2015): 17–18, https://doi.org/10.1097/01.JAA.0000460925.13380.92; and A. Grover, J. M. Orlowski, and C. E. Erikson, "The Nation's Physician Workforce and Future Challenges," *American Journal of the Medical Sciences* 352, no. 1 (2016): 11–19, https://doi.org/10.1016/j.amjms.2015.10.009.

34. G. Sánchez, T. Nevarez, W. Schink, and D. E. Hayes-Bautista, "Latino Physicians in the United States, 1980–2010: A Thirty-Year Overview from the Censuses," *Academic Medicine* 90, no. 7 (2015): 906–12, https://doi.org/10.1097/ACM.0000000000000619; L. M. Hamel, R. Chapman, M. Malloy, et al., "Critical Shortage of African American Medical Oncologists in the United States," *Journal of Clinical Oncology* 33, no. 32 (2015): 3697–700, https://ascopubs.org/doi/abs/10.1200/JCO.2014.59.2493.

35. See, for example, J. L. Graves, "Biological V. Social Definitions of Race: Implications for Modern Biomedical Research," *Review of Black Political Economy* 37, no. 1 (2009): 43–60, https://doi.org/10.1007/s12114-009-9053-3; and J. L. Graves, C. Reiber, M. Hurtado, A. Thanukos, and T. Wolpaw, "Evolutionary Science as a Method to Facilitate Higher Level Thinking and Reasoning in Medical Training," *Evolution, Medicine, & Public Health* no. 1 (January 2016): 358–68, https://doi.org/10.1093/emph/eow029.

36. R. Jones, *White Too Long: The Legacy of White Supremacy in American Christianity* (New York: Simon & Schuster, 2020).

37. See, for example, E. H. Ecklund, *Why Science and Faith Need Each Other: Eight Shared Values That Move Us Beyond Fear* (Grand Rapids, MO: Brazos, 2020).

38. See "Our Mission," BioLogos, accessed May 4, 2021, https://biologos.org/about-us#our-mission; also see Joe's discussion, "The Genetics of Race (Part 1)," https://biologos.org/podcast-episodes/joseph-graves-the-genetics-of-race-part-1, and "The Genetics of Race (Part 2)," https://biologos.org/podcast-episodes/joseph-graves-the-genetics-of-race-part-2.

39. J. Barndt, *Becoming the Anti-Racist Church: Journeying Toward Wholeness* (Minneapolis: Fortress, 2011).

40. P. H. Collins and S. Bilge, *Intersectionality*, 2nd ed. (Cambridge, UK: Polity Press, 2020).

41. *Status of Women in the States*, https://statusofwomendata.org/explore-the-data/employment-and-earnings/employment-and-earnings/#eefigure2.3.

42. A. Dinno, "Homicide Rates of Transgender Individuals in the United States: 2010–2014," *American Journal of Public Health* 107, no. 9 (2017): 1441–47, https://doi.org/10.2105/AJPH.2017.303878.

CONCLUSIONS

1. N. Wade, *A Troublesome Inheritance: Genes, Race, and Human History* (New York: Penguin, 2014).

2. C. Darwin, *Narrative of the Surveying Voyages of the* Adventure *and* Beagle, vol. III (London: Henry Colburn, 1839).

3. S. J. Olshansky, T. Antonucci, L. Berkman, R. H. Binstock, A. Boersch-Supan, J. T. Cacioppo, . . . and J. Rowe. "Differences in Life Expectancy due to Race and Educational Differences Are Widening, and Many May Not Catch Up," *Health Affairs* 31, no. 8 (2012): 1803–13.

4. T. Andrasfay and N. Goldman, "Reductions in 2020 US Life Expectancy Due to COVID-19 and the Disproportionate Impact on the Black and Latino Populations," *Proceedings of the National Academy of Sciences* 118, no. 5 (2021): e2014746118, https://doi.org/10.1073/pnas.2014746118.

INDEX